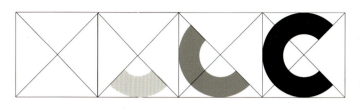

2011
COMPETITIONS Annual

2011年度竞赛合集

G·斯坦利·科利尔（博士，美国建筑师协会荣誉会员）编 常文心 译

辽宁科学技术出版社

图书在版编目（CIP）数据

2011年度竞赛合集 / （美）科利尔（Collyer, G.S.）
编；常文心译. -- 沈阳：辽宁科学技术出版社，2012.8
　　ISBN 978-7-5381-7478-6

　　Ⅰ．①2… Ⅱ．①科… ②常… Ⅲ．①建筑设计－作
品集－世界－现代 Ⅳ．①TU206

　　中国版本图书馆CIP数据核字(2012)第089060号

出版发行：辽宁科学技术出版社
　　　　　（地址：沈阳市和平区十一纬路29号　邮编：110003）
印　刷　者：利丰雅高印刷（深圳）有限公司
经　销　者：各地新华书店
幅面尺寸：267mm×267mm
印　　张：15
字　　数：50千字
印　　数：1～1500
出版时间：2012年 8 月第 1 版
印刷时间：2012年 8 月第 1 次印刷
责任编辑：陈慈良　常　雪
封面设计：吴　杨
版式设计：吴　杨
责任校对：周　文
书　　号：ISBN 978-7-5381-7478-6
定　　价：228.00元

联系电话：024-23284360
邮购热线：024-23284502
E-mail: lnkjc@126.com
http://www.lnkj.com.cn
本书网址：www.lnkj.cn/uri.sh/7478

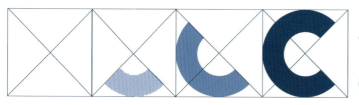

2011
COMPETITIONS Annual

2011年度竞赛合集

Editor in Chief: G. Stanley Collyer, Ph.D, Hon. AIA
Associate Editor: Daniel Madryga, MS Arch.
总编：G·斯坦利·科利尔（博士，美国建筑师协会荣誉会员）
助理编辑：丹尼尔·马德里加（建筑师）
特别感谢：威廉·摩根；特德·桑德斯特拉；埃里克·戈登伯格
With contributions by **William Morgan, Ted Sandstra,**
and Eric Goldemberg

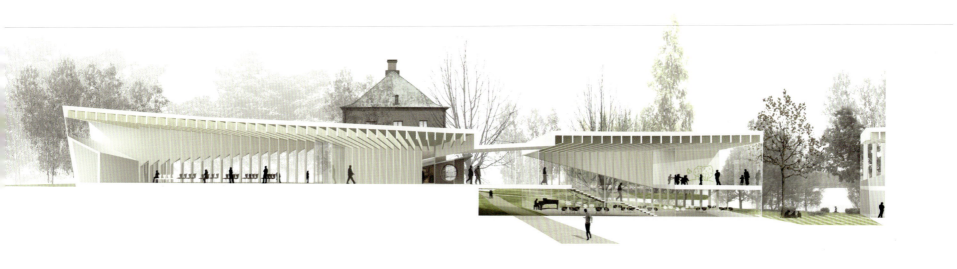

Contents

目录

Introduction 简介

这本《2011年度竞赛合集》呈现了过去一年中
比较令人关注的竞赛设计作品，大多数作品可
以在我们的网站www.competitions.org找到。
它呈现了各种开放、匿名的竞赛以及限制、邀
请竞赛，后者的入围名单经过了严格的筛选流
程。由于经济的滞后，美国几乎没有真实项目
的竞赛，大多数竞赛都设有报价和最终挑选候
选人流程。相反，我们发现在欧洲和亚洲，大
量的限制和开放式竞赛正呈增长趋势。最令人
惊讶的是塞尔维亚和阿尔巴尼亚的竞赛，尽管
曾经拥有外国公司建造的项目，这两个国家直
至最近才对外国人开放竞赛。在东欧，波兰近
几年一直在进行国内竞赛，最近也开始对国际
设计人才开放了具有高知名度的竞赛。
但是纵观全球，最大的新闻发生在与中国大陆
一海峡之隔的台湾岛。最近，台湾的竞赛项
目主要以开放式的两阶段竞赛为主。这让中等
规模的设计公司可以与普利兹克奖得主公平竞
争。这还涌现出大量平时出现在高知名度竞赛
中具有试验性的设计。纽约建筑公司雷泽+乌
梅莫托在美国没有获得大型项目，却在高雄港
和游船中心竞赛和台北流行音乐文化中心竞赛
中蝉联。帮派工作室在后者中获得的第二名，
表明竞赛中有一些美国高排名公司参与竞争，
其中包括d/A工作室和墨菲西斯事务所。在中
国大陆，又是另一番景象，大型跨国公司在各
个竞赛中都占据主导地位。在试验性竞赛中，
具有创新精神的事务所大受欢迎，如OMA、
斯蒂文·霍尔、赫尔佐格和德梅隆等。更加透
明的竞赛通常由外国政府派遣的公司获胜，例
如：美国使馆竞赛由SOM获胜。
由于越来越多的竞赛采用邀请式，小型事务所
和独立设计在竞赛中慢慢消失。这与网络的普
及导致了概念竞赛的大爆发，它们其中的一些
很合理，另一些则毫无意义。后者的产生主要
是因为设计师想与同行看齐，以吸引赞助商、
评委的注意，进入正式竞赛。
本书感谢众多建筑师、竞赛顾问和组织单位，
特别是辛辛那提大学的设计、建筑、艺术和规
划学院。
斯坦利·科利尔

This 2011 COMPETITIONS Annual represents some of the more interesting competitions which have taken place over the past year, most of which have been covered on our website, www.competitions.org. It represents a mix of open, anonymous competitions, and restricted, invited competitions, whereby finalists in the latter have been shortlisted via a screening process. Due mainly to a lagging economy, competitions for real projects were almost absent in the U.S., and most of those that did take place almost invariably involved an RfQ and shortlisting process. Instead, we saw a growing list of both limited and open competitions in Europe and Asia. Most surprising were competitions taking place in Serbia and Albania, two countries which had not opened competitions to foreigners until very recently—previously commissioned projects to foreign firms notwithstanding. Also in Eastern Europe, Poland, a country which has been staging their own internal competitions for several years, has opened some high-profile competitions to the international community of design professionals.

But the big news on the global scene has been what has been happening on the island of Taiwan, just off the coast of mainland China. Recently, major projects there have been the subject of open, two-stage competitions. This has allowed medium-sized firms to participate on an equal footing with Pritzker prize winners. It has also allowed for more experimentation than one would normally see when high-profile commissions are at stake. Reiser + Umemoto, a New York-based firm which has not received that many major commissions in the U.S., won the Kaohsiung Port and Cruise Service Center Terminal Competition not long after winning the Taipei Pop Music Cultural Center competition in Taipei. Studio Gang's second place in the latter competition marked the top ranking of several U.S. firms in that contest, including Office d/A and Morphosis.

On the Chinese mainland, it has been a different story, where large firms with offices in several countries have dominated the competition scene. When experimentation has taken place, it normally has been by firms with an innovative reputation: Office of Metropolitan Architecture (OMA), Steven Holl, Herzog de Meuron, etc. The more transparent competitions there, although usually invited, have normally been those administered by foreign governments for their own legations, i.e., the American embassy competition won by SOM.

Since so many competitions are now by invitation, participation by small firms and individual architects has declined. This, as well as universal access to the internet, has led to the explosion of ideas competitions, some legitimate, others quite frivolous. As to the latter, the suspicion arises that some are simply looking to line their pockets at the expense of the profession. Thus, it is always incumbent on those wishing to enter competitions to scrutinize the sponsors, as well as the jury.

As with any work of this magnitude we are indebted to many architects, competition advisers and institutions, not the least of which is the College of Design, Architecture, Art and Planning at the University of Cincinnati.

Stanley Collyer

L'entrée actuelle de l'école d'architecture s'avère comme inadaptée par ses dimensions, sa localité et ses expressions pour recevoir un visiteur.
Le projet propose donc une entrée qui se fait dans la cour. Elle devient donc le **premier lieu d'accueil**.
Une fois passé la porte cochère, le visiteur est guidé par une pente qui le dirige vers la nouvelle entrée principale ensoleillée par le soleil du sud. Les locaux de l'entrée, spacieux et lumineux, correspondent aux besoins contemporains et mènent directement vers les lieux les plus publiques c'est à dire à l'auditorium et à la salle d'exposition.
Aussi une liaison directe avec la circulation verticale est facilement accessible.
Les escaliers de **l'auditorium profitent du dénivellement** du terrain.
La cour de l'ancien séminaire est aujourd'hui rarement utilisée

par les étudiants de l'école, faute de manque d'accès à cette dernière et en raison de fonctions trop spécifiques du rez-de-chaussée actuel comme des salles de cours.
Les caractéristiques typiques d'un séminaire reposent entre autre sur la typologie de la cour fermée.
Le projet propose donc de venir **fermer la cour**, tout en s'appliquant sur les anciens plans du séminaire où cette fermeture était prévue. Elle garantie une **réappropriation de la cour** par les étudiants. Elle est englobée par des fonctions publiques et semi-publiques qui font de la cour et du nouveau rez-de-chaussée un **grand lieu de rassemblement**. La cour est entièrement entourée par les parois vitrées du nouvel emblème qui renforcent les vues et l'accessibilité de celle-ci.
Le projet s'implante comme une ligne légère tout au long de la cour.

8

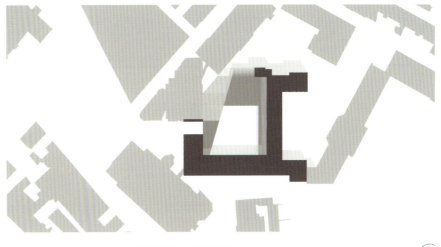

| atelier de bois | FONCTIONS |
| cafétéria | SEMI-PUBLIQUES |

entrée	
auditorium	
salle d'exposition	FONCTIONS PUBLIQUES

 la circulation se ferme

PLAN DU SITE | échelle 1.1000

Embarcation as the Ultimate Experience

当乘船成为终极体验

SITE

The New Kaohsiung Port and Cruise Terminal in Taiwan 台湾新高雄港和游船中心竞赛

Background

高雄港口
优胜者
宗迈建筑师事务所/
雷泽+乌梅莫托 RUR建筑事务所
台湾/纽约
上图：场地鸟瞰图
对页：港口开放区

**Kaohsiung Port Terminal
Winner**
Fei & Cheng Associates/
**Reiser+Umemoto RUR
Architecture**
Taiwan/New York City
ABOVE
Birdseye view of site
OPPOSITE PAGE
Landside view of terminal

Many buildings in close proximity to bodies of water seem to have that joie de vivre about them. Whether it is Sea Ranch, The Bilbao Guggenheim, Oslo Opera House or summer residences in the Hamptons, the proximity of water somehow manages to stimulate designers to produce excitement in a relaxed atmosphere.

From the Greek temples to Spas in England, construction of major structures on oceans and rivers was always more likely to reflect modern trends in architecture, rather than simply replicating a style from the past. Recent waterfront projects such as the Yokohama International Port Terminal—a competition won by Foreign Office Architects—and Canada Place in Vancouver are examples of cities recognizing the need to push the envelope when redesigning port terminal facilities. And so it was with the results of the Kaohsiung Port and Cruise Service Center competition .

Not only is Kaohsiung a major port facility on the island, it is seen as a major terminal for future water transit to the Chinese mainland. The goal of the competition was to identify a design that will enhance the travel experience of passengers, make it a principal departure destination for cruise ships, and provide recreational opportunities for the local populace. Moreover, it is understood that the new facility should add to the urban vitality of the immediate vicinity.

许多水边建筑似乎都受益匪浅。无论是海洋牧场、毕尔巴鄂·古根海姆、奥斯陆歌剧院还是汉普斯顿的夏日别墅，亲水性都激发了设计打造放松氛围的灵感。从希腊神庙到英格兰温泉浴场，海洋和河流上的主题结构总是能反映建筑的现代潮流，而不是简单地复制过去的风格。新近的滨水项目，如横滨国际港口码头（由外国建筑事务所取得竞赛胜利）和温哥华的加拿大广场都是城市认识到重新设计港口设施的范例。这也是高雄港和游船中心竞赛的主旨。

高雄不仅是台湾岛一个重要的港口设施，而且是未来通往中国大陆的水上交通枢纽。竞赛的目标是打造一个能够提升游客旅行体验的设计，使其成为游船的主要离港码头，并且为当地民众提供休闲娱乐的机会。此外，竞赛意识到新设施应该增添其周边区域的城市跃度。

优胜设计

关注台湾流行音乐中心竞赛的人们马上就会发现宗迈建筑师事务所和雷泽+乌梅莫托 RUR建筑事务所的获胜设计与他们参与那个竞赛的作品的共同之处。两个项目所采用的相似的曲线塔楼设计不容忽视。除了这个强烈的象征表述之外，优胜设计也有许多创新。设计的道路系统和室内深受评审好评："内部空间和功能区的设置简洁明晰。特别是主要室内空间的流畅性为使用者带来了当地公共建筑少有的高端空间体验。'木板路'装饰了高雄的港口区域和海边空间，通过一系列城市滨水设施与周边的城市环境结合在一起。"尽管关于建筑朝向城市的硬立面还有质疑，该结构的角色是作为"海洋和城市之间的临界面"——奥雅纳工程顾问公司的介入让设计团队对施工更加自信："RUR、奥雅纳和宗迈建筑师事务所共同打造了具有非凡专业经验的卓越团队。设计现实可行，充分考虑了预算、结构和施工。"——编者

'The combination of RUR, ARUP and Chang & Fei makes an excellent team with exceptional experience and expertise.'

"RUR、奥雅纳和宗迈建筑师事务所共同打造了具有非凡专业经验的卓越团队。"

The Winning Design

Anyone following the recent Pop Music Center Competition in Taiwan would have immediately recognized the resemblance between the winning design in that competition by **Fei & Cheng Associates with Reiser+Umemoto RUR Architecture**, and their team entry here. In both cases the similarities in the curvilinear tower design could hardly be ignored. Aside from that strong symbolic statement, the winning design had a lot going for it. It's circulation plan and interior got high marks from the jury:

"The internal spatial and functional arrangement is simple and explicit. Specifically, the flow and fluidity of the main interior spaces is to offer a high-quality spatial experience rare in local public buildings. The 'boardwalk' created furnishes Kaohsiung's port area with important seaside open space, which can be integrated with the surrounding urban spaces into a series of waterfront amenities for the city." Although there was a question concerning the hard facade of the building facing the city—"the role of the structure as an "interface between ocean and city"—the inclusion of Arup as an asset on the team had to be a confidence builder: "The combination of RUR, ARUP and Chang & Fei makes an excellent team with exceptional experience and expertise. The proposal is both realistic and feasible with regards to budget, structure and construction." -Ed

高雄港竞赛优胜作品
宗迈建筑师事务所/
雷泽+乌梅莫托 RUR建筑事务所
台湾/纽约
上图
木板路夜景
左图
室内远景
对页，左上
入境（上）和出境（下）道路形式说明
对页，右上
大堂/售票区和安全区
对页，左中
地面层平面图
对页，右中
纵切面
对页，左下
主层平面图
对页，右下
三层平面图

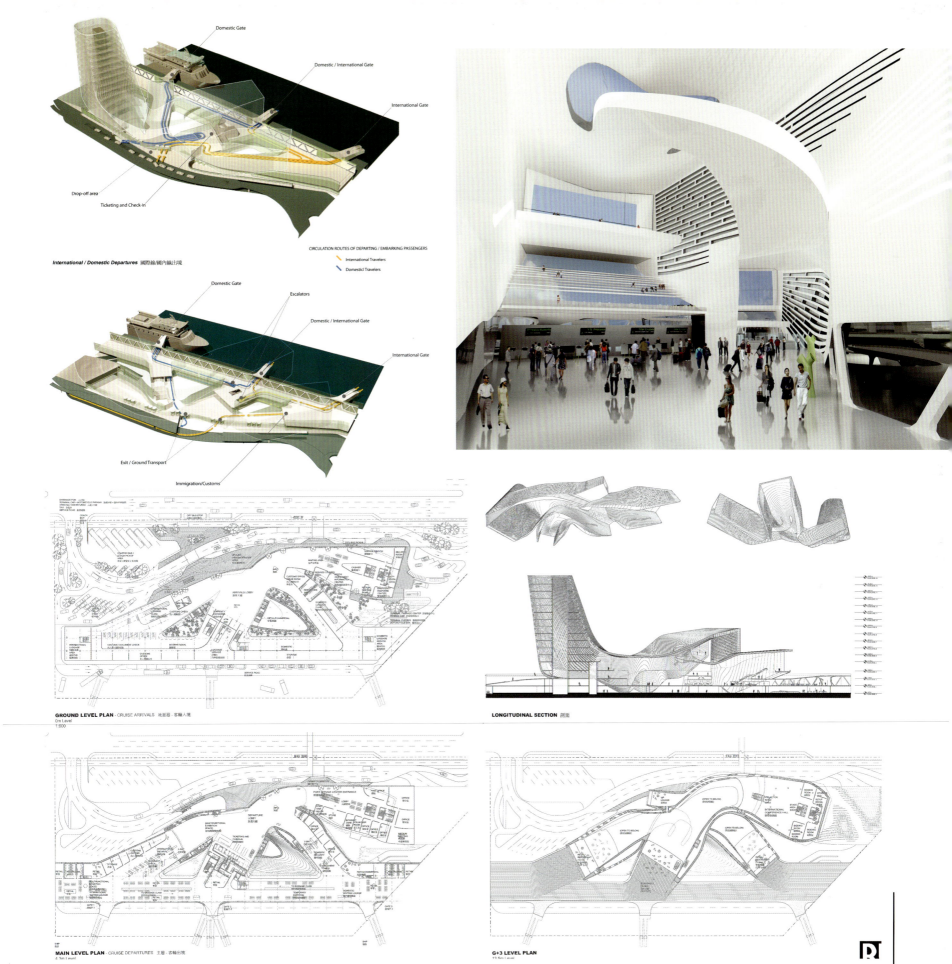

International / Domestic Departures 國際線/國內線/出境

Domestic Gate
Domestic / International Gate
International Gate
Drop-off area
Ticketing and Check-in

CIRCULATION ROUTES OF DEPARTING / EMBARKING PASSENGERS
International Travelers
Domestic Travelers

Domestic Gate
Escalators
Domestic / International Gate
International Gate
Exit / Ground Transport
Immigration/Customs

GROUND LEVEL PLAN - CRUISE ARRIVALS 地面層·客輪入境
0m Level
1:800

LONGITUDINAL SECTION 剖面

MAIN LEVEL PLAN - CRUISE DEPARTURES 主層·客輪出境
4.7m Level

G+3 LEVEL PLAN
13.5m Level

高雄港竞赛优胜作品
宗迈建筑师事务所/
雷泽+乌梅莫托 RUR建筑事务所
台湾/纽约
左图
侧面远景
左下
模型鸟瞰图
右下
入口
对页
功能区分解说明

Kaohsiung Port Terminal Winner

Fei & Cheng Associates/ Reiser+Umemoto RUR Architecture

Taiwan/New York City

LEFT
Landside perspective
BELOW, LEFT
Birdseye view of model
BELOW, RIGHT
Approach to entrance
OPPOSITE PAGE
Exploded illustration of functions

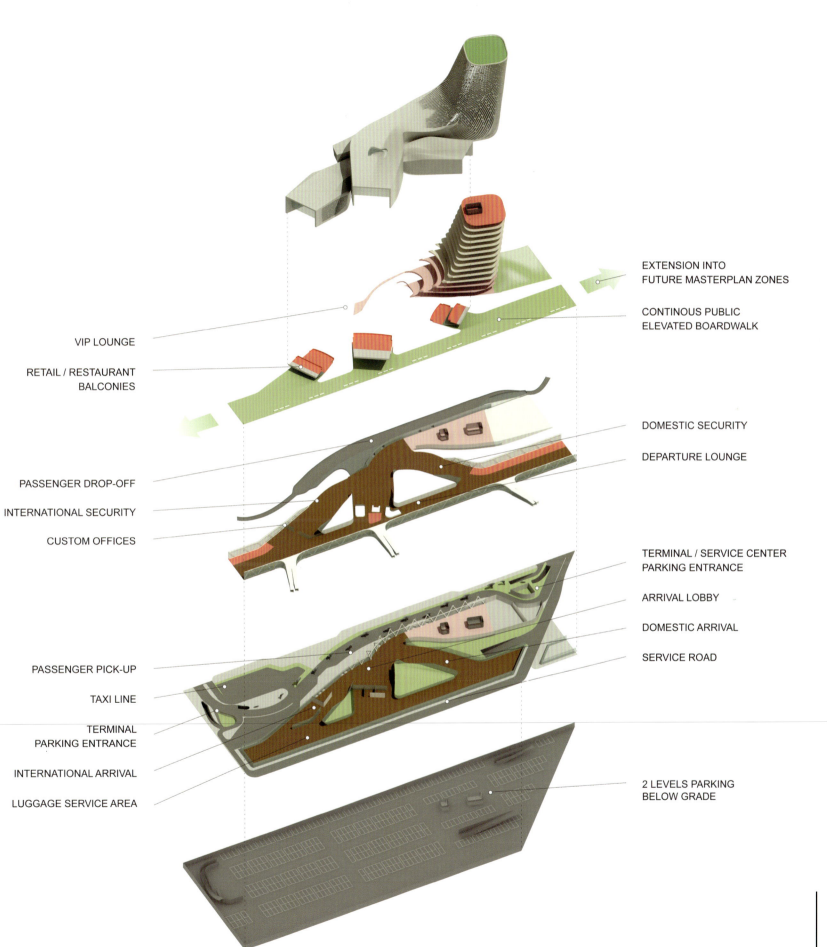

EXTENSION INTO
FUTURE MASTERPLAN ZONES

CONTINOUS PUBLIC
ELEVATED BOARDWALK

VIP LOUNGE

RETAIL / RESTAURANT
BALCONIES

DOMESTIC SECURITY

DEPARTURE LOUNGE

PASSENGER DROP-OFF

INTERNATIONAL SECURITY

CUSTOM OFFICES

TERMINAL / SERVICE CENTER
PARKING ENTRANCE

ARRIVAL LOBBY

DOMESTIC ARRIVAL

SERVICE ROAD

PASSENGER PICK-UP

TAXI LINE

TERMINAL
PARKING ENTRANCE

INTERNATIONAL ARRIVAL

LUGGAGE SERVICE AREA

2 LEVELS PARKING
BELOW GRADE

Second Prize

Asymptote Architecture

New York, NY
with Artech Architects/Kris Yao, Taiwan
The "elegant" **Asymptote** proposal may have only failed due to the two-tower idea. According to jurors, "The terminal space covered with an innovative, shell-like roof structure is distinctive and beautiful," and "the open public space below also makes this scheme successful as a marine gateway." The fact that the main building "establishes a good relationship between the urban blocks and the waters" probably pushed this design to the forefront of the adjudication process. As always, there were differing takes by the jury on the architec-

tural expression of the scheme. According to one, "the design style is minimalistic and understated, but as a national gateway may be somewhat lacking in intensity."

Finally, there was a concern about separating the office tower into two buildings, "creating an enclosing gesture," and the treatment of the plaza space in relationship with the terminal function. All the jurors seemed to concur on this, that "the east office building blocking the view from the cruise arrival axis" was a serious consideration in the final evaluation. Still, this entry would seem to be the most poetic in terms of architectural expression. -Ed

二等奖
渐近线建筑事务所（纽约）
大元建筑及设计事务所/姚仁喜（台湾）
渐近线事务所"优雅"的设计的失败可能只在于"双子塔"的理念。评审称："港口空间上方的创新、壳状屋顶结构独特而美观"，"下方的开放式公共空间让设计成为了一个成功的航海门户。"事实上，主楼"在城市空间和水体之间所打造的良好的关系"是该设计获得评审好评的关键。和往常一样，评审们在建筑表达形式上意见不同。根据一位评审所说："设计风格简洁为低调，但是作为一个国际性门户设施，它缺乏强烈感。"
最终，将办公楼一分为二也是一个问题，这"形成了闭合空间"。设计还将广场处理为港口功能相通。评审们一致认为"东侧的办公楼阻碍了游船到达轴线的视野"，这个问题十分严重。无论怎样，这个作品是建筑表达中最具诗意的设计。——编者

城市指標 / *URBAN ICON*

HAMI RASHID手稿 / CONCEPT SKETCHES BY HAMI RASHID

公共空間 / PUBLIC PLAZA

OPPOSITE PAGE
Birdseye view of terminal from landside
LEFT
Pedestrian perspective from plaza
BELOW
Arrival/Departure area

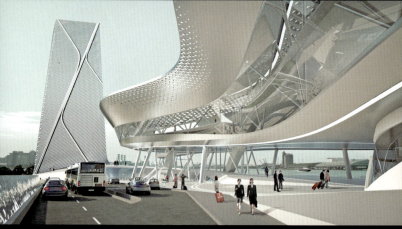

建築設計構想及功能分佈 / *CIRCULATION & PROGRAM*

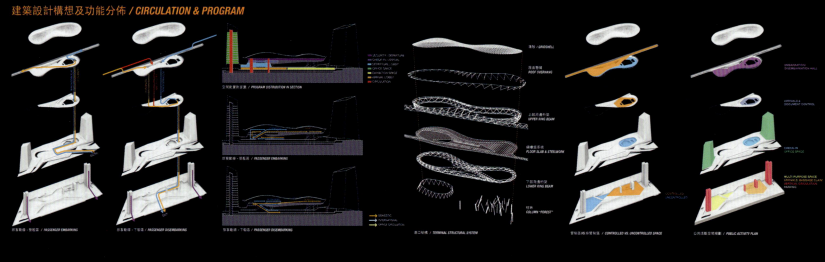

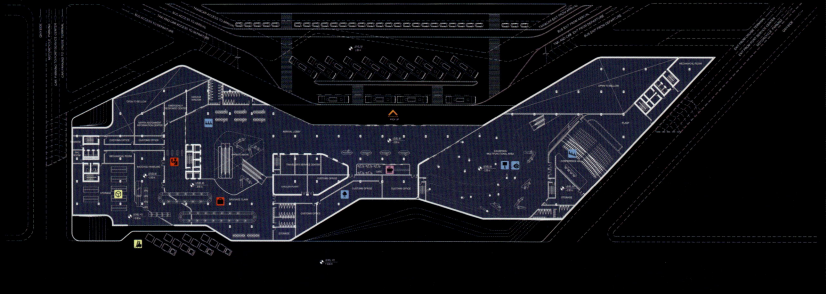

入境與行李提領樓層 / *ARRIVAL & BAGGAGE LEVEL, - 2.5M*

一等奖

渐近线建筑事务所（纽约）
大元建筑及设计事务所/姚仁喜（台湾）
对页
室内景象、下客区、组织/功能区、平面图
下图
城市远景、海滨景观、剖面

Second Prize

Asymptote Architecture, New York, NY
with Artech Architects/Kris Yao, Taiwan

OPPOSITE PAGE
Interior view, drop off area, organization/program, and floor plan
BELOW
Urban perspective, seaside view, section

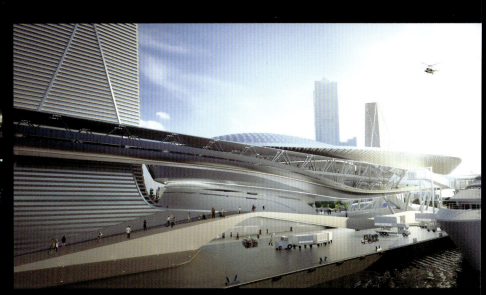

纵向剖面图 / **LONG SECTION**

12/12 高雄海洋門户
KAOHSIUNG MARINE GATEWAY

Third Prize
Ricky Liu & Associates Architects+Planners, Taiwan
With Takenaka Corporation/Masahiro Morita, Japan

The entry by **Ricky Liu & Associates** won high marks on several counts from almost all the jurors. The site plan, its relationship to the urban context and "beautiful building form" were common observations. One found it "superior to the other schemes in terms of overall formal aesthetic and the geometric structural shell. Another observed, "the most important part of the spatial design is the plaza space open to the city, but possible confusion between the controlled zones, arrival and departure circulation and the public areas is a concern." Although the "beehive-shape" design was found to be especially interesting, there was an overriding budgetary concern: "But how will this structure realistically transform into the building skin, how will it be maintained, and under what scope will it be built?" This critique was almost universal among the jurors, and it no doubt contributed to its lower ranking. Could it be that the Sydney Opera House budgetary debacle was lurking in the back of the juror's minds? -Ed

RIGHT, ABOVE
Waterside pedestrian perspective
RIGHT, BELOW
Aerial view of site

三等奖
刘培森建筑师事务所（台湾）
竹中工程公司/森田正广（日本）

刘培森建筑师事务所的设计几乎获得了全体评审的好评。场地规划、与城市环境的联系和"美观的建筑形式"是建筑的公认的优点。"设计在整体造型美学和几何结构外壳的设计上优于其他方案。""空间设计中最重要的部分就是朝向城市的广场空间，但是控制区、出入境道路交通和公共区域的可能性混乱令人担忧。"尽管"蜂窝形"设计十分有趣，却能造成过高的预算："这个结构如何实现建筑外壳的转变？如何维护？怎样建造？"评审们几乎都持有这种论调，毫无疑问，这降低了设计的评分。或许评审们正在担忧设计像悉尼歌剧院一样拥有庞大的预算。——编者

右上
水畔走道远景
右下
场地鸟瞰图

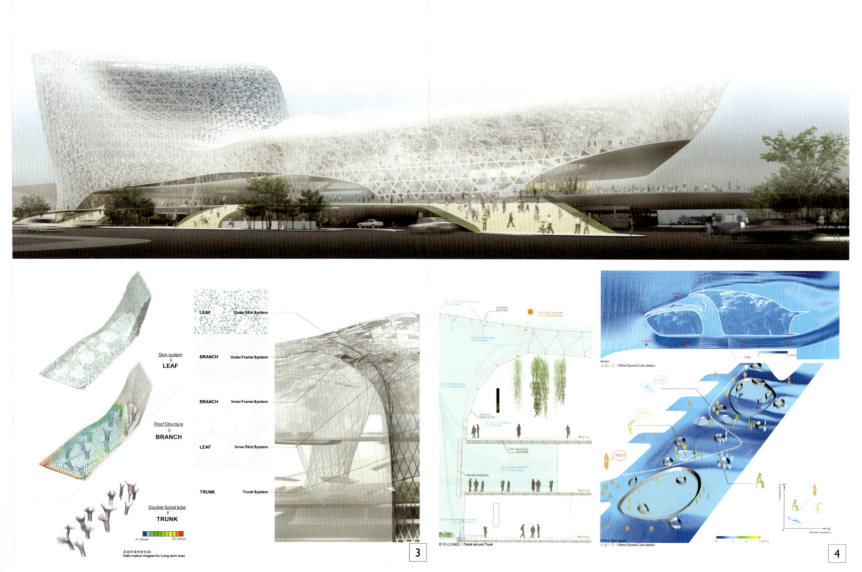

Skin system
= **LEAF**

Roof Structure
= **BRANCH**

Double Spiral tube
= **TRUNK**

長期荷重変形和図
Deformation diagram for Long-term load

LEAF — Outer Skin System

BRANCH — Outer Frame System

BRANCH — Inner Frame System

LEAF — Inner Skin System

TRUNK — Trunk System

幹柱 切詳細図 / Detail around Trunk

Section
风速計算 / Wind Speed Calculation

开放空间风速計算 / Wind Speed Calculation

3

4

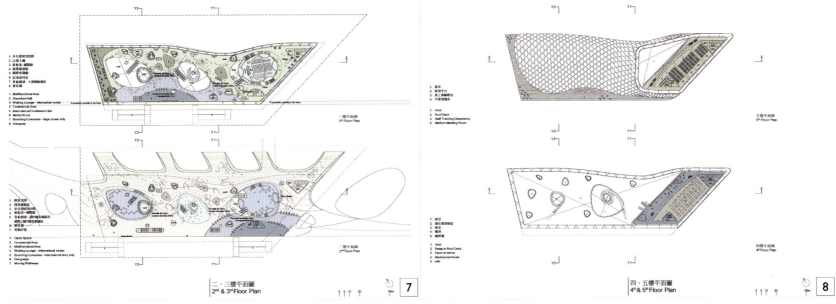

二、三樓平面圖
2nd & 3rd Floor Plan

四、五樓平面圖
4th & 5th Floor Plan

7

8

地面層平面圖
Ground Floor Plan

Arrival Area
1. Arrival Lobby
2. Traveller's Service Center
3. Baggage Handling Area
4. Baggage Claim - controlled zone
5. Arrival Hall - controlled zone
6. Customs Offices
7. Storage

Departures Area
8. Departure Lobby
9. Check - in Counter &
   Baggage Check-in
10. Temporary Bus Parking
11. Temporary Coach Parking

Office Area
12. Entrance Lobby
13. Service Lobby
14. Service Counter
15. Security Room
16. Volunteer Room
17. Post Office
18. Convenience Store
19. Service Window
20. Cashier
21. Berth Assignment Information
    Center
22. Emergency Response Center
23. Internet Zone
24. Waiting Area (for passenger and office)
24. Dock Operation Area

ABOVE
Approach from east
OPPOSITE PAGE, ABOVE
Birdseye view
OPPOSITE PAGE, BELOW
Landside perspective

上图
东侧入口
对页，上图
鸟瞰图
对页，下图
公共区远景

Honorable Mention – I

JET Architecture Inc./Edward Kim, Canada
with CXT Architects Inc./Dan Teh, Canada and
Archasia Design Group/Sao-You, Taiwan

JET Architecture/Edward Kim from Canada proposed a spiral rampway as a major aesthetic device and organizational concept. "The rampway leading to the departure level dictates the spatial design of this proposal and at the same time dominates the building elevations. However, this functional architectural element falls short of transforming itself into a built space that offers aesthetic quality and a pleasing experience." Another juror put it this way: "This is a scheme that presents an open public plaza protected by an office building up in the air, supported by a structure. It creates wonderful outdoor opportunities for the public. The scheme has handled the departure/arrival circulation in a very articulate manner. However, bringing cars up to the fourth level creates various issues in the organization of the building. Also, a critical point brought up in the jury process was that the terminal space, which is supposed to offer the passengers a special experience, is fitted into a rather conventional structure in the podium." -Ed

荣誉奖1：

喷射建筑公司/爱德华·金（加拿大）
CXT建筑公司/丹·泰（加拿大）和瀚亚设计/徐少游（台湾）

来自加拿大的喷射建筑公司/爱德华·金设计了一个螺旋坡道作为项目的主要美学特征和组织概念。"通往离境层的坡道显示了方案的空间设计，同时也占据了建筑立面的支配地位。然而，这一功能建筑元素的不足在于无法转换为优美舒适的建筑空间。"另一位评审则认为："这一方案呈现了一个由办公楼保护的开放式公共广场。它为公众打造了完美的户外活动空间。设计还以清晰的方式解决了出入境道路交通问题。然而，将停车场设在四层为建筑的组织结构带来了一系列问题。此外，评审过程中一个重要的点就是港口空间的设计。这个空间需要为绿卡带来独特的体验，融入一个相对传统的底座结构之中。"——编者

Honorable Mention II

HMC Group Inc. / Raymond Pan
Los Angeles, CA
with HOY Architects &
Associates/Charles Hsueh, Taiwan

The **HMC Group** led by Raymond Pan allowed nature to play a big role in their design: "They presented an in-depth exploration of green building issues and translated it into convincing results. The building volume responds sensitively to the solar path. In other words, the form has been sculpted by sunlight. The roof design attempts to amalgamate the entire landscape while creating a series of open space for the public. However, the key requirement of a memorable, charismatic national gateway may not have been given quite enough emphasis." The jury evaluation of this scheme was probably summed up best by Juror A: "This unique scheme demonstrates a clear articulation and organization of the program. It presents several interesting ideas to make the building feasible and ecological. It is regrettable that it is missing a strong partie to tie all the ideas together." -Ed

LEFT, ABOVE
View from bay
LEFT, BELOW
View to lobby from above
OPPOSITE, ABOVE
Aerial view from seaside
OPPOSITE, BELOW
Arrival perspective

荣誉奖2：

HMC集团/雷蒙德·潘（洛杉矶）
华业建筑师事务所/薛昭信（台湾）

由雷蒙德·潘领衔的HMC集团让自然在设计中起到了重要的作用："他们呈现了对绿色建筑的深度探索，并将其运用到设计之中。建筑造型灵敏地反映了太阳路径。也就是说，建筑造型是由阳光雕刻而出的。屋顶设计视图合并整个景观，同时为公众提供开放空间。然而，方案并没有充分重视竞赛的主要需求——打造一个富有魅力的壮丽门户空间。"评审A的话最能总结评审团的看法："这个独特的设计展示了清晰的连接和组织结构。若干个有趣的概念体现了建筑的可行性和生态型。遗憾的是，它缺乏一个将各个理念串联在一起的重要环节。"

Bucolic Site as Museum Context

by William Morgan 博物馆的田园风情

威廉·摩根

The Serlachius Museum Competition in Finland 芬兰希尔拉切丝博物馆竞赛

Background

In Finland, a land where architectural competitions are a way of life, a design contest for an addition to a small art museum drew the greatest number of entries in Finnish competition history. That the Serlachius Museum in an out-of-the way city could attract 579 entrants from 41 countries may say something about the flat world economy. But it is more likely a measure of the attractiveness of the project, the reputation of the client, and the above-board way competitions are run in Finland. It also says something about this Nordic country's expectations about the quality of life and the high esteem in which cultural institutions are held. The art of architecture, too, is deemed essential to the fabric of national life, and it is not just the exclusive province of large cities or major corporations. Mänttä is the 91st largest municipality in Finland, situated in the forest halfway between Jyväskylä and Tampere, yet size and apparent obscurity do not diminish the demand for architectural greatness.

As is with many mill towns in Finland, Mänttä was dominated by a single company whose owners took a benevolent interest in all aspects of the lives of the employees. The Serlachius family were discerning collectors, primarily of Finnish art. So, the factory owner's widow, Ruth Serlachius, opened the family's collection to the public in 1945, turning over Joenniemi Manor (1930, Jarl Eklund, architect) to the Gösta Serlachius Fine Arts Foundation.

The museum was greatly enlarged in 2000 when the company headquarters building was added to the Serlachius Museum. Originally called the "White House," this flat-roofed, stark Functionalist building of 1934 couldn't be more different from the grand Gustavian manor next door. Demonstrating the avant-garde sensibilities of the Serlachius family, this cool piece of Modernism was designed by Walter and Bertel Jung, the latter an associate of Eliel Saarinen, as well as Helsinki's first city planner.

> That the Serlachius Museum in an out-of-the way city could attract 579 entrants from 41 countries may not just say something about the attractiveness of the project, but also the reputation of the client, and the above-board way competitions are run in Finland.

But the new addition would be part of the manor that is set in one of Finland's rare examples of an English park-cum-garden, with commanding view down to Lake Melasjärvi. The new 15 million, 3,000 square-meter block will add a main entrance lobby, conference rooms, a restaurant, offices, and collection storage, as well as exhibition spaces. But "the natural connection of the proposed extension with the unique environment, and how well the solution is in harmony with the environmental values of the manor grounds," was a key criterion for evaluation. And it is fair to say that those entries that lacked understanding of the landscape issues were the first to be eliminated.

Even more important was the program's brief that the new building must be "a high-quality representation of modern construction" and "provide the area with a new attraction whose values will stand the test of time." In other words, the "idea behind the competition is that the new building shall be a work of art, not simply a functional framework for operations." The jury included museum director Pauli Sivonen, the city architect, and three other architects. Henrik de la Chapelle, chairman of the museum's board, led the jury. The competition rules were those of the Finnish Association of Architects. Demonstrating the museum's serious commitment to the design, top prizes were 40,000, 30,000, and 20,000, along with a commitment to erect the winning scheme. There were also three purchase prizes of 10,000 each, and eight honorable mentions.

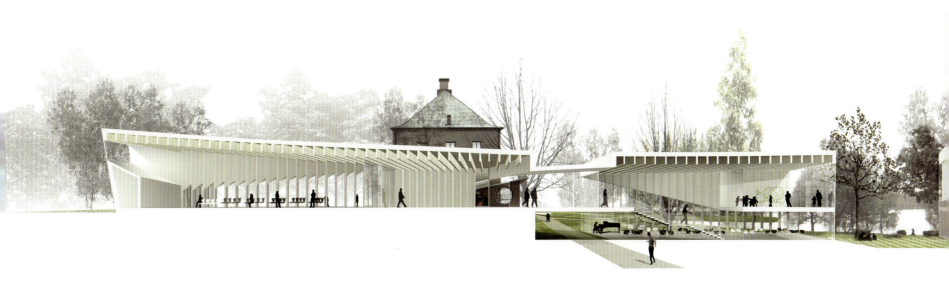

背景

在芬兰这个以建筑竞赛为生活方式之一的国度，一座小型艺术博物馆的扩建设计竞赛吸引了芬兰竞赛史上最多的参赛者。这个位于不知名小城的希尔拉切丝博物馆竞赛吸引了来自41个国家的579位参赛者，这充分说明了全球经济一体化。项目的吸引力、委托人的声望和芬兰公开的竞赛形式都是竞赛如此受欢迎的原因。同时，这也体现了北欧国家对生活质量的期望值和对自身文化设施的尊重。建筑的艺术是国家生活的重要组成部分，不仅仅是大城市和大企业的专利。曼塔在芬兰城市中排名91位，位于捷瓦斯基拉和坦佩雷之间的一片森林之中，然而小规模和默默无闻并不影响其对伟大建筑的追求。

与芬兰的诸多工业城镇一样，曼塔被一家公司所统治。这家公司重视员工生活的方方面面。希尔拉切丝家族是颇具眼光的收藏家，主要收藏芬兰艺术品。因此，工厂主的遗孀——露丝·希尔拉切丝在1945年向公众开放了家族收藏，将其转交给乔涅米·马诺尔（1930，贾尔·艾克兰，建筑师）所掌管的戈斯塔·希尔拉切丝艺术基金会。博物馆在2000年进行了扩建，将公司总部大楼添加到希尔拉切丝博物馆之中。这座建于1934年的朴实无华的平顶建筑与旁边的加斯塔维安庄园大相径庭，曾被称作"白房子"。为了体现希尔拉切丝家族的先锋敏锐感，这座出色的现代主义建筑由沃尔特和波特尔·荣格（伊利尔·沙里宁的助手，也是赫尔辛基的首位城市规划师）建造。但是新建的结构采用了芬兰罕见的英式花园风格，俯瞰着下方的麦拉斯贾维湖。造价1,500万欧元、占地3,000平方米的建筑结构将增添一个主入口大厅、若干会议室、一家餐厅、许多办公室、一个收藏仓库和若干展览空间。但是"新建结构与独特环境的自然联系以及如何使设计与庄园和谐统一"是竞赛评价的重点。那些欠缺景观理解的作品被首先淘汰。

更重要的是项目要求新建筑必须"高度展示现代建筑结构"并"提供经得起时间考验的全新魅力"。换句话说，"竞赛背后的理念是新建筑应该是一件艺术品，而不仅仅是一座运营机构"。竞赛评审包括博物馆主管保利·西沃农、城市建筑师和三名其他的建筑师。竞赛规则由芬兰建筑师协会制定。为了体现博物馆对设计的重视，最高奖金分别为40,000欧元、30,000欧元和20,000欧元，并且优胜设计将获得博物馆的建造权。此外，竞赛还设有三个10,000欧元的参与奖和八个荣誉奖。

一等奖
"平行线"

MX_SI建筑工作室（西班牙，巴塞罗那）
团队成员
埃克托·门多萨
马拉·帕尔迪达
鲍里斯·贝赞

上图
剖面图
对页
从广场看项目

Overall, the museum sought thoughtful design, with respect for the landscape and for the museum's needs–the jurors were looking for quality, not flash.

总而言之，博物馆寻求经过深思熟虑的设计，既要尊重景观设计，又要满足博物馆的需求——评审需要的是质量，而不是灵光一现。

First Place
"Parallels"

MX_SI Architectural Studio
Barcelona, Spain
Team
Héctor Mendoza
Mara Partida,
Boris Bezan

ABOVE
Section
OPPOSITE PAGE
View from plaza

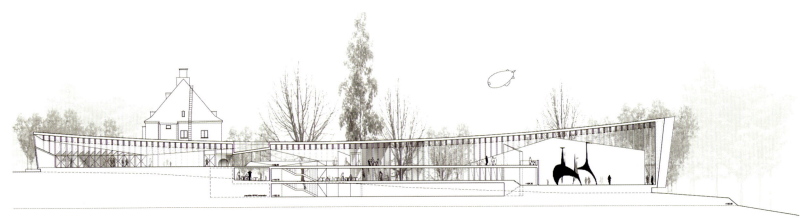

First Place *"Parallels"*
MX_SI Architectural Studio, Barcelona, Spain

The winning design by the Barcelona-based firm of MX_SI Architectural Studio, whose principals include two Mexicans, Héctor Mendoza and Mara Partida, and Slovenian Boris Bezan, was a splendid choice. Wrapped in a delicate wooden building, the museum seems genuinely quiet and in harmony with the landscape. Yet the roof gently sloping up at both ends and the vertical siding beautifully reproduces the rhythm of the trees in the surrounding landscape. The walls do not reveal that the interior plan is composed of several angles, allowing "a spatial experience wherein the indoor and outdoor spaces are in continuous dialogue." The new museum is sited in such a way that would allow further extensions, something not possible with the more sculptural proposals. In an almost immodest comment from Finns, MX_SI's design is praised for its demonstration of its knowledge of "the existing building stock, the history of the location and the client, Finnish culture, and contemporary architecture."

MX_SI建筑工作室（西班牙，巴塞罗那）
由巴塞罗那公司MX_SI建筑工作室提交的优胜设计十分耀眼。工作室由两名墨西哥总监埃克托·门多萨和马拉·帕尔迪达和斯洛文尼亚人鲍里斯·贝赞组成。博物馆被设在精致的木制建筑之中，看起来宁静平和，与景观和谐相处。然而，两端缓缓上升的屋顶和垂直外墙完美地再现了周边树木的韵律。墙壁不仅展示的由多重角度组成的室内设计，还"在室内外空间之间形成了连续的对话和空间体验"。新博物馆的设计便于进一步扩张，比其他作品更具灵活性。芬兰的评委充分肯定了MX_SI的设计，称其完美融合了"原有的建筑结构、项目地点和委托人的历史、芬兰文化和现代建筑"。

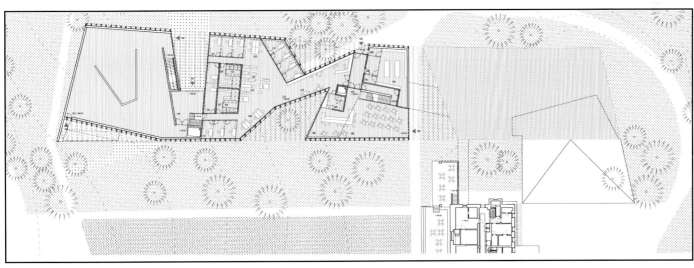

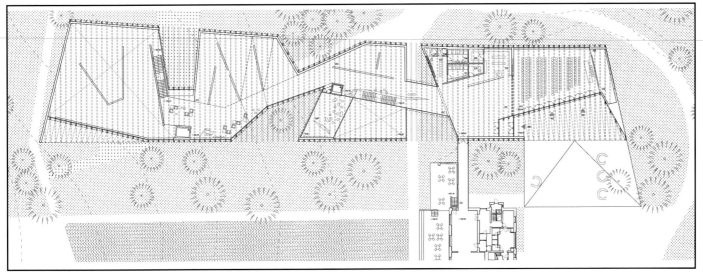

ABOVE
Site plan
LEFT, ABOVE
Ground level floor plan
LEFT, BELOW
1st level floor plan

OPPOSITE PAGE, ABOVE
Exhibit hall
OPPOSITE, MIDDLE
Section
OPPOSITE, BELOW LEFT
Pedestrian perspective, model
OPPOSITE, BELOW RIGHT
Aerial view of model

上图
总规划图
左上
底层平面图
左下
二层平面图
对页，上图
展览厅
对页，中图
剖面图
对页，左下
模型人行道远景
对页，右下
模型鸟瞰图

Second Place: *"Thyra"*
Mikko Heikkinen and Markku Komonen
Helsinki, Finland

The runner up design is a magnificent piece of modernism by leading Finnish architects, Mikko Heikkinen and Markku Komonen. Employing their typical dry wit, they called their entry Thyra (shoebox). This "splendid proposal is convincing with its clarity and beautiful exhibition rooms." Recalling museum designs by Heikkinen and Komonen (the shoebox is one of their signatures), this "a classic piece of modern architecture: plain on the outside but rich on the inside. After feasting on all kinds of shapes, one returns to the basics." The jury called Thyra "a fresh deviation from the mainstream of the competition." The basic black box, however, would have been enlivened with glass ramps on the park side.   -WM

二等奖："塞拉"
米克·海基宁和马尔库·科莫宁（芬兰，赫尔辛基）

亚军设计是由芬兰建筑师米克·海基宁和马尔库·科莫宁所设计的宏伟的现代主义作品。他们运用了自己一贯敏锐的蔡司，将作品命名为"塞拉"（鞋盒）。这件"璀璨的作品一起简洁而美观的展览室而令人信服"。这件作品体现了米克·海基宁和马尔库·科莫宁典型的博物馆设计（鞋盒结构是他们的鲜明特色之一），"是一件典型现代建筑：外观平凡，内部丰富。在尝试了多种造型之后，人们开始回归最初。"评审称，塞拉是"主流竞赛中一抹新鲜偏差"。然而，如果在朝向公园一侧增添玻璃坡道，基本的黑盒结构将增色不少。——威廉·摩根

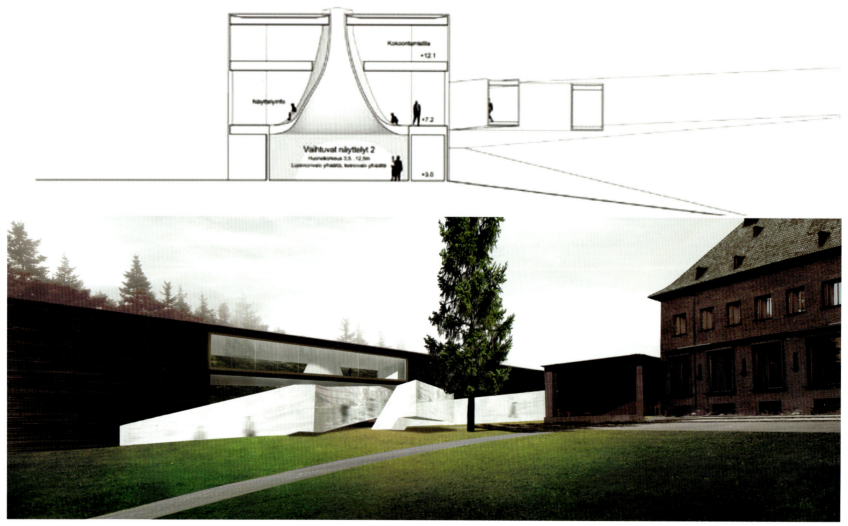

左图
室内组织结构
下图
立面图和剖面图
对页，上图
室内景象
对页，中图
光轴剖面图
对页，下图
入口

LEFT
View of interior organization
BELOW
Elevation and sections
OPPOSITE PAGE, ABOVE
Interior view
OPPOSITE PAGE, MIDDLE
section showing light shaft
OPPOSITE, BELOW
View to entrance

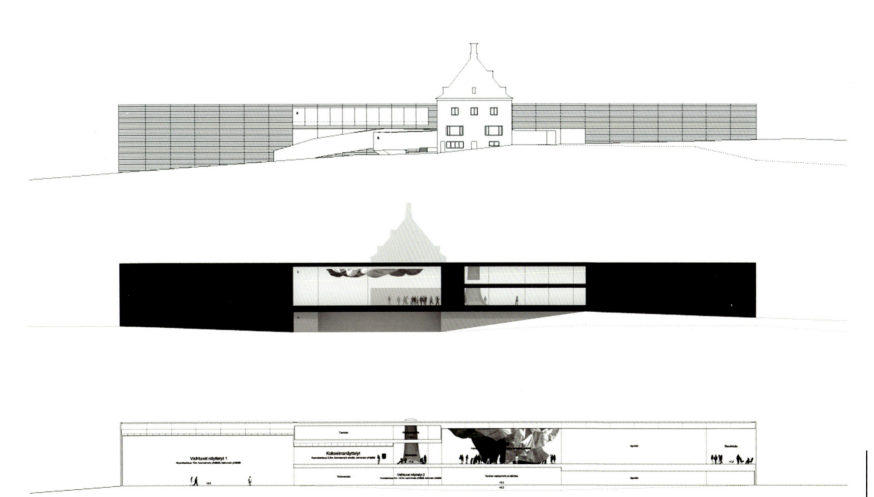

Second Place: *"Thyra"*
Mikko Heikkinen and Markku Komonen
Helsinki, Finland

ABOVE	上图
Interior perspective	室内远景
LEFT	左图
View to entrance	入口
BELOW	下图
Exhibition floor plan	展览区平面图
	对页
OPPOSITE PAGE	总规划图
Site plan	

二等奖："塞拉"
米克·海基宁和马尔库·科莫宁（芬兰，赫尔辛基）

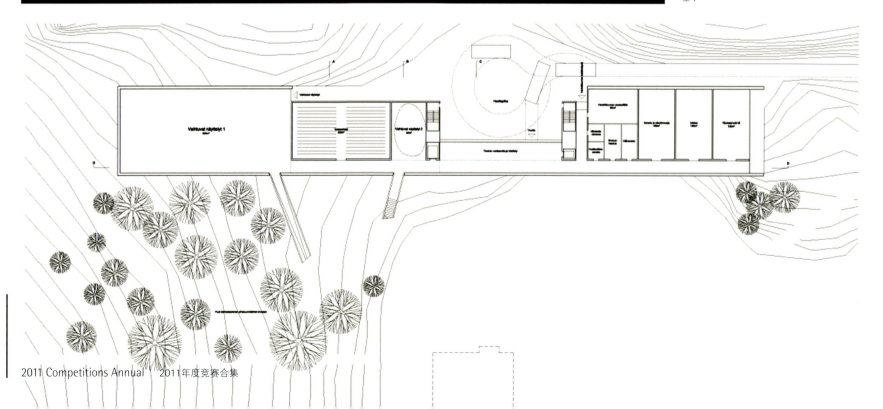

Huoltopiha

Huoltosisäänkäynti
▽
+3.0

+15.7

Ravintolan sisäänkäynti
△

Ulkonäyttelyalue

+7.5

Gösta Serlachius ◁
+25.5

Taavetinsaari

Sauna

Pysäköintialue

三等奖："露丝·S"
利库和卡特里·荣卡（芬兰，赫尔辛基）

三等奖项目"露丝·S"（纪念了博物馆创始人露丝·希尔拉切丝）由赫尔辛基建筑师利库和卡特里·荣卡设计，是最激进的作品。设计颇具雕塑感的金字塔造型"像露丝的凉帽一样轻盈"，评审评论说："屋顶造型设计与众不同、精致而令人愉悦。"设计具有鲜明的瑞典建筑特色（例如，勒·柯布西耶的费尔米尼教堂）。灵活、轻松而具有完好的功能价值，露丝·S是一件抽象的设计，尽管可能永远不会被实现，却还是深深吸引着评审委员会。——威廉·摩根

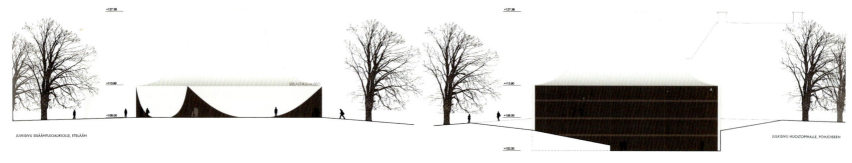

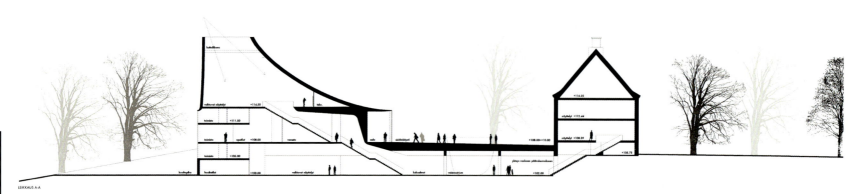

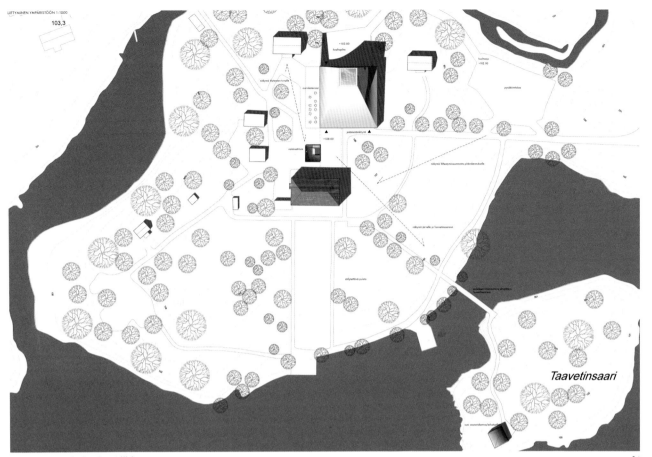

2.6 Serlachius-museen Gösten laajennus · yleinen kansainvälinen suunnittelukilpailu

Taavetinsaari

LIITTYMINEN YMPÄRISTÖÖN 1:1000
103,3

Rut

Third Place: "*Ruth S*"
Riku and Katri Rönka
Helsinki

The third place finisher, *Ruth S.* (sweetly honoring museum founder Ruth Serlachius) by Helsinki architects Riku and Katri Rönka is the most radical. Calling its sculptural pyramidal shape "as light and airy as Ruth's summer hat," the jury notes the "play on the forms of the roof is distinctive, delicate, and delightful," with clear precedents in Swiss architecture (i.e., Le Corbusier, say, his Firminy church). Clever, light-hearted, and functionally well planned, Ruth S. is something of a sleeper—the sort of way-out design that appeals to a jury knowing that it will probably never be considered as a serious possibility for this site. -WM

THIS PAGE
Aerial view of site (top) and elevations (bottom)
OPPOSITE PAGE, ABOVE
Pedestrian perspective (left) and Interior view (right)
OPPOSITE, BELOW
Section

JULKISIVU LÄHESTYMISSUUNTAAN, ITÄÄN

JULKISIVUMATERIAALIT:

本页
场地鸟瞰图（上）和立面图（下）
对页，上图
人行道（左）和室内（右）
对页，下图
剖面图

JULKISIVU LÄNTEEN

Purchase Prize 1

"Piparminttu"
Thomas Gebert
Sankt Gallen, Switzerland

BELOW
Aerial view of model

参与奖1

"皮帕敏图"
托马斯·吉伯特（瑞士，圣加仑）

下图
模型鸟瞰图

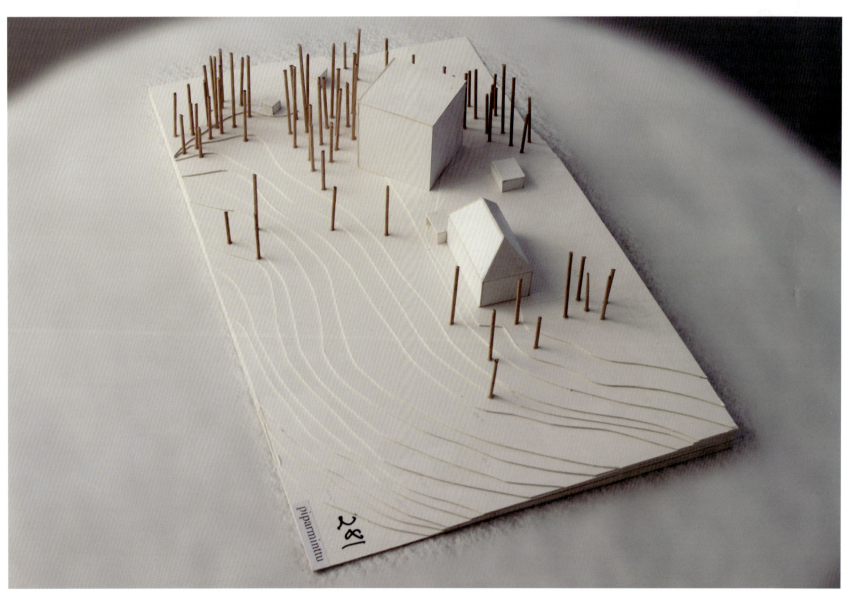

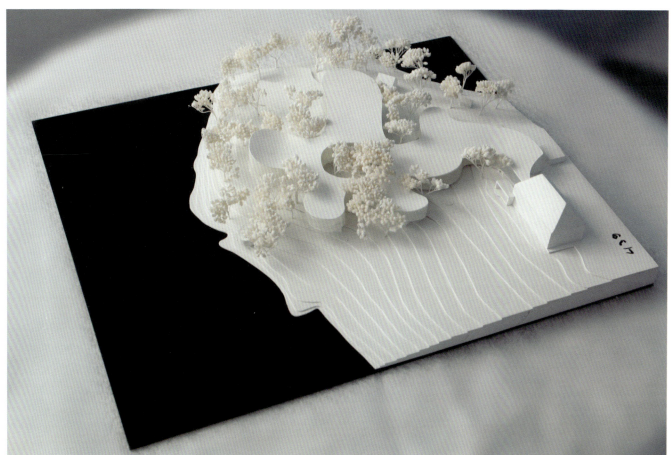

Purchase Prize 2

"HP1002"
magma architecture
Ostermann & Kleinheinz
Berlin, Germany

LEFT
Aerial view of model

参与奖2

"HP1002"
岩浆建筑事务所
奥斯特尔曼&克林辛兹（德国，柏林）

左图
模型鸟瞰图

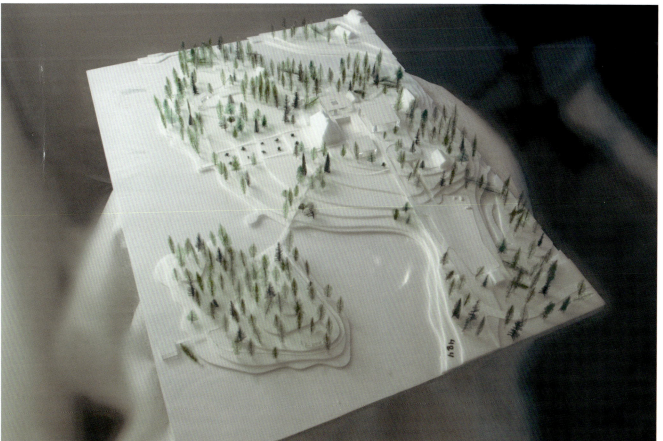

Purchase Prize 3

"MKS101"
MAKS | Architecture &
Urbanism
The Netherlands

LEFT
Aerial view of model

参与奖3

"MKS101"
MAKS建筑规划事务所（荷兰）

左图
模型鸟瞰图

Restoring and Reinventing Albanian Identity

by Dan Madryga

重塑阿尔巴尼亚形象　地拉那和宗教和谐清真寺和博物馆

丹·马德里加

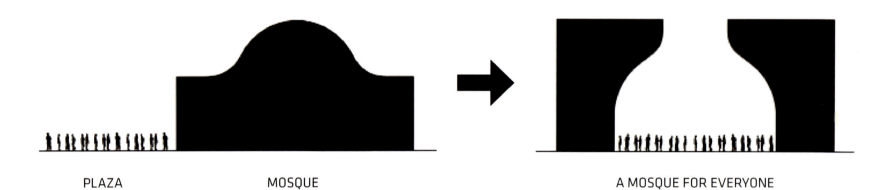

PLAZA　　　　　　　MOSQUE　　　　　　　　　　　　　A MOSQUE FOR EVERYONE

A New Mosque and Museum of Tirana & Religious Harmony

Tirana, Albania might be the last place that many would associate with cutting edge architecture. The capital of a poor country still struggling to sweep away the lingering vestiges of the communist era, it is understandable that architecture and design have not always been a top priority. Yet in the face of the city's struggles, Tirana is striving to reclaim and reshape its image and identity, and international design competitions are playing no small role in this movement. And while Tirana has yet to be associated with contemporary architecture, the implementation of these design competitions has introduced a handful of renowned architecture firms to the city with high hopes of bolstering the international image of Albania.

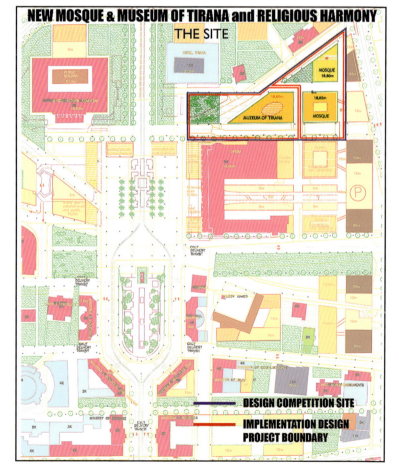

人们也许永远不会想到阿尔巴尼亚的地拉那会和先锋建筑联系在一起。这个贫困国家的首都仍然在贫困线上挣扎，因此，建筑和设计显然不是首要考虑的因素。然而，地拉那正努力重塑自身的形象和身份，国际设计竞赛在此次活动中起到了举足轻重的作用。尽管地拉那还没有和现代建筑联系在一起，这些设计竞赛已经为城市引入了大量知名建筑事务所，也高度提升了阿尔巴尼亚的国际形象。

因此，竞赛对建筑的规模有所要求：宏大的清真寺必须能够在平日接纳1,000名朝圣者，周五接纳5,000名朝圣者，盛大的节日则需要接纳10,000名朝圣者。作为清真寺的附属建筑，项目还要求设计一个伊斯兰文化中心，用于教学研究，其内部将设有图书馆、多功能厅和研讨教室。

竞赛的另一个元素是地拉那宗教和谐博物馆。博物馆将各个宗教和背景的市民汇聚到一起。除了展现地拉那的历史之外，博物馆将以城市宗教遗产为中心，突出体现宗教和谐。教育公众更加了解伊斯兰文化并提升宗教容忍度是一个高尚的目标。

项目位于斯坎德贝格广场，地理位置突出了它的重要性。作为地拉那的行政文化中心，广场周边汇集了主要政府建筑和大量的博物馆和剧院。广场本身就是2003年的一个设计竞赛的产物，重塑了城市中心（虽然目前这里仍是一片混乱），使其成为更人性化的步行街区。清真寺和文化中心设在歌剧院和地拉那酒店旁的一片三角形场地上，将成为地拉那城市景观中耀眼的元素。

这次两阶段的国际竞赛由地拉那市政府和阿尔巴尼亚穆斯林教会联合举办，由耐瓦特·赛因和阿尔

坦·何塞担任顾问。

一百多支参赛队伍（大多数来自欧洲）向竞赛提交了参赛作品。三月初，候选委员会选择了五支队伍获得45,000欧元的奖金，他们分别为：

BIG——丹麦，哥本哈根

seARCH——荷兰，阿姆斯特丹

扎哈·哈迪德建筑事务所——英国，伦敦

安德里亚斯·佩雷亚和NEXO——西班牙，马德里

建筑工作室——法国，巴黎

设计作品经过一个多样化的欧洲评审团评价：

埃迪·拉马——地拉那市长（阿尔巴尼亚）

保罗·波姆——建筑师（德国，科隆）

维德兰·米米卡——建筑师，贝尔拉格学院院长（克罗地亚）

彼得·斯威尼——51N4E事务所合伙人和建筑师（布鲁塞尔）

恩佐·西维埃罗教授——工程师，威尼斯建筑大学教授（威尼斯）

阿尔坦·史克莱里——建筑师（阿尔巴尼亚，地拉那）

西克力·莱利——穆斯林教会代表

5月1日，评审委员会宣布BIG为竞赛优胜者。

BIG的设计"所有人的清真寺"发源于一个简单而有效的设计理念。项目起源于两个主轴扭曲的美感设置：以斯坎德贝格广场为中心的城市网格和清真寺必须朝向麦加的主墙。护墙与墙壁上部与街面对齐，地面层则采取了扭曲的朝向，将主广场朝向穆斯林圣地。BIG的托马斯·克里斯托弗尔森解释了这一设计："将基准线朝向麦加解决了总体规划中的两难境地：清真寺的三角形布局在角落中有些拖拉；现在让它朝向广场，两都是邻里建筑。合成的建筑（无论是清真寺本身，还是周边的半穹顶空间）让人想起了传统伊斯兰建筑的拱顶和拱门。"

总体规划和布局产生了一系列广场：清真寺两侧各有一座小广场，前面还有一个以尖塔为标志的大型集会空间。这些广场的设置帮助容纳了宗教节日期间的大规模朝圣者，也为平时的使用提供了充足的空间。建筑的外立面上排列着精致的方形小窗，这一设计灵感来自于伊斯兰的传统屏风。玻璃装配图案与弧型墙壁的组合将在圣殿内形成千变万化的光影效果。

Hence emphasis in the brief concerning the size of the building: a grand mosque that can adequately serve 1000 prayers on normal days, 5000 on Fridays, and up to 10,000 during holy feasts. Supporting this mosque, the program also specifies the design of a Center of Islamic Culture that will house teaching, learning, and research facilities including a library, multipurpose hall, and seminar classrooms.

Another component of the competition program, the Museum of Tirana and Religious Harmony, moves beyond the realm of the Muslim community in an explicit gesture to bring together citizens from all faiths and backgrounds. Aside from presenting the general history of Tirana, the museum will focus on the city's religious heritage, highlighting both the turbulent moment of suppression under communism as well as the religious harmony that has since been reinstated. Educating the public about Islamic culture and promoting religious tolerance at a time when relations between religious communities are strained throughout the world is certainly a noble objective.

Underlining the importance of this project is its prominent site on Scanderbeg Square, the administrative and cultural center of Tirana where major government buildings share an expansive public space with museums and theaters. The square itself was the subject of a 2003 design competition that will eventually reclaim the urban center—at present a rather chaotic vehicular hub—as a pedestrian zone with a more human scale. Situated on triangular site adjacent to the Opera and Hotel Tirana, the Mosque and Cultural Center will be a highly visible component of Tirana's urban landscape.

First Place

BIG I Bjarke Ingels Group
Copenhagen, Denmark

ABOVE
Birdseye view
OPPOSITE, ABOVE
Fundamental approach to problem
OPPOSITE, BELOW
Site plan

一等奖

BIG（丹麦，哥本哈根）

上图
鸟瞰图
对页，上图
解决问题的基本方案
对页，下图
总规划图

The two-stage, international competition was organized by the City of Tirana and the Albanian Muslim community and advised by Nevat Sayin and Artan Hysa.

Over one hundred teams—the vast majority European—submitted qualifications for the first stage. In early March, the short-listing committee selected five teams to receive an honorarium of 45,000 Euros each to develop designs:

- **Bjarke Ingels Group I BIG** – Copenhagen, Denmark
- **seARCH** – Amsterdam, Holland
- **Zaha Hadid Architects** – London, UK
- **Andreas Perea Ortega with NEXO** – Madrid, Spain
- **Architecture Studio** – Paris, France

The designs were judged by a diverse European panel:

- Edi Rama - Mayor of Tirana, Albania
- Paul Boehm – architect, Cologne, Germany
- Vedran Mimica – Croatian architect; current director of the Berlage Institute
- Peter Swinnen – Partner and architect at 51N4E, Brussels
- Prof. Enzo Siviero – engineer; Professor at University IUAV, Venice
- Artan Shkreli – architect, Tirana, Albania• Shyqyri Rreli – Muslim community representative

On May 1st the panel announced **Bjarke Ingels Group (BIG)** as the winner.

The singular form of **BIG'**s design, "A Mosque for All," is driven by a simple yet effective concept. The project derives its warped aesthetic through the accommodation of two significant axes: the city grid that conforms to Scanderbeg Square and the necessary orientation of the Mosque's main wall to face Mecca. While the parapet and upper portions of the walls align to the street, the ground level twists to orient itself and the main plaza towards the Muslim Holy City. Thomas

'BIG's design is a success in its ability to address and merge a number of contextual and cultural concerns.'

Christoffersen of BIG explains this key decision: "The alignment towards Mecca solves the dilemma inherent in the master plan: in its triangular layout the mosque was somehow tugged in the corner; now it sits at the end of the plaza, framed by its two neighbors. The resultant architecture evokes the curved domes and arches of traditional Islamic architecture, for both the mosque itself and the semi-domed spaces around it."

A series of plazas result from the master plan's layout and massing: two small plazas on either side of mosque and one large gathering space in front punctuated by a minaret. These plazas are arranged to help accommodate the large groups of worshippers during holy days while also providing compelling spaces for more secular daily use. The facades of the buildings are riddled with a fine pattern of small rectangular windows inspired by Islamic mashrabiya screens. This pattern of glazing coupled with the curving walls will allow for ever-changing light patterns within the worship hall.

BIG's design is a success in its ability to address and merge a number of contextual and cultural concerns. As Mayor Rama noted: "The winning proposal was chosen for its ability to create an inviting public space flexible enough to accommodate daily users and large religious events, while harmonically connecting with the Scanderbeg square, the city of Tirana and its citizens across different religions." -DM

First Place

BIG I Bjarke Ingels Group
Copenhagen, Denmark

ABOVE
Plaza as prayer venue
LEFT
Plaza view
BELOW
Participant view

OPPOSITE, ABOVE
Daytime view of plaza
OPPOSITE, BELOW
Plaza at night

一等奖

BIG（丹麦，哥本哈根）

上图
朝圣者广场
左图
广场
下图
朝圣者

对页，上图
广场日景
对页，下图
广场夜景

First Place

BIG I Bjarke Ingels Group

Copenhagen, Denmark

Partners-in-Charge:
Bjarke Ingels
Thomas Christoffersen
Project Leader: Leon Rost

Project Team
Marcella Martinez, Se Yoon Park, Alessandro Ronfini, Daniel Kidd, Julian Nin Liang, Erick Kristanto, Ho Kyung Lee

Collaborators:
Martha Schwartz Landscape
Buro Happold
Speirs & Major
Lutzenberger & Lutzenberger,
Global Cultural Asset Management
Glessner Group

一等奖

BIG (丹麦，哥本哈根)

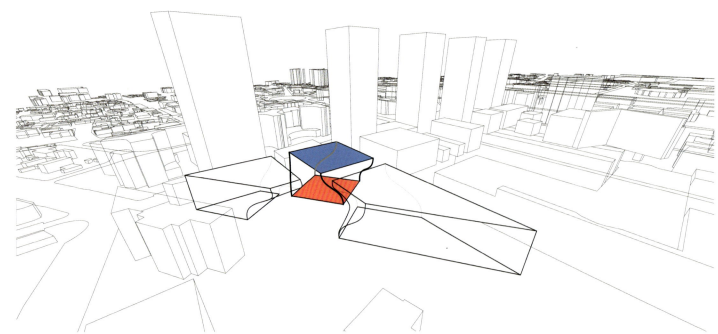

一等奖

BIG（丹麦，哥本哈根）

左图
从广场看博物馆
下图
图书馆
对页，上图
清真寺室内
对页，下图
场地在城市环境中的示意图，清真寺为彩色标记

Winning Design

BIG I Bjarke Ingels Group
Copenhagen, Denmark

LEFT
View of museum from plaza
BELOW
Library perspective
OPPOSITE, ABOVE
Mosque interior
OPPOSITE, BELOW
Site in urban context with mosque in color

Finalist 入围奖

seARCH
with

Lola Landscape Architects,
Daan Roosegaarde en Pieters
Bouwtechniek

seARCH和洛拉景观建筑事务所、丹·罗斯贾德和皮特斯·布泰尼克（荷兰，阿姆斯特丹）

design team
Bjarne Mastenbroek, Kathrin Hanf,
Peter Veenstra,
With Andrea Verdecchia, Pablo
Domingo, Teresa Avella, Simona
Schroder, Manuel Granados, José
Rodriguez
Amsterdam, The Netherlands

LEFT　　　　　　　　　　　左图
View to entrance　　　　　入口
BELOW　　　　　　　　　　下图
Plaza　　　　　　　　　　　广场
OPPOSITE PAGE　　　　　　对页
Celebraion venue　　　　　节庆场合

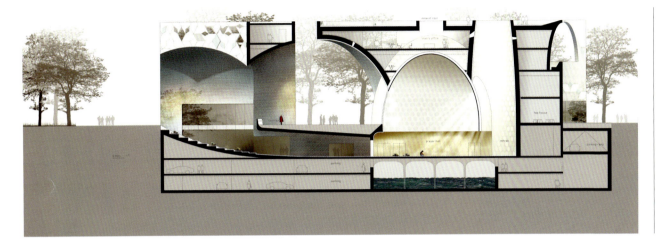

Finalist

seARCH with Lola Landscape
Architects, Daan Roosegaarde en
Pieters Bouwtechniek
Amsterdam, The Netherlands

LEFT	左图
Section	剖面图
BELOW	下图
Foyer	门厅
OPPOSITE PAGE, ABOVE	对页，上图
View to northeast	项目东北角
OPPOSITE, BELOW	对页，下图
aerial view of site	场地鸟瞰图

入围奖

seARCH和洛拉景观建筑事务所、丹·罗斯贾德和
皮特斯·布泰尼克（荷兰，阿姆斯特丹）

左图
清真寺内部
下图
博物馆大厅
对页，上图
模型鸟瞰图
对页，下图
清真寺平面图

Finalist

seARCH with Lola Landscape Architects,
Daan Roosegaarde en Pieters Bouwtechniek
Amsterdam, The Netherlands

LEFT
Mosque interior
BELOW
Museum hall
OPPOSITE ABOVE
Aerial view of model
OPPOSITE, BELOW
Mosque floor plans

Finalist

seARCH /Lola Landscape Architects, Daan Roosegaarde en Pieters Bouwtechniek
Amsterdam, The Netherlands

Specifically aiming to avoid a "bird's eye view" architectural solution in favor of the experiential, Amsterdam's seARCH generated a design that reveals itself gradually but rewardingly. The museum for example, is a carefully designed journey. From the below grade lobby accessed from a sunken square, visitors take escalators towards the exhibition spaces housed in a dramatically cantilevered volume. There they are presented with a panoramic window that looks out over Scanderbeg Square.

The architecture of the mosque places a clear priority on the interior experience. Appearing as a stylized yet unassuming cube from the exterior street facades, the entrance approach from the partially sheltered main plaza reveals a series of shimmering, irregular domed ceilings carved out of the rectangular volumes. These domes, along with an interspersed grove of trees, give the plaza and mosque interior an intimate spatial quality despite the grand scale. While this inside-out approach does diminish the mosque's external presence from certain street-side vantage points, seARCH's entry offers one of the more spiritually intriguing worship spaces of the finalists. -DM

入围奖
seARCH和洛拉景观建筑事务所、丹·罗斯贾德和皮特斯·布泰尼克（荷兰，阿姆斯特丹）
来自阿姆斯特丹的seARCH特别避免了"鸟瞰式"建筑形式，他们的设计逐步显露自己的优势。博物馆被精心设计成一次特别的旅程。游客通过下沉广场进入地下大堂，乘坐电梯进入悬臂结构中的展览空间。呈现在他们面前的是斯坎德贝格广场的全景。

清真寺的设计以室内空间为主。从街面上看，建筑低调而朴实，由主广场进入的入口处呈现出各式不规则的穹顶。这些穹顶在树木的点缀下，为广场和清真寺内部带来亲密的空间体验。——丹·马德里加

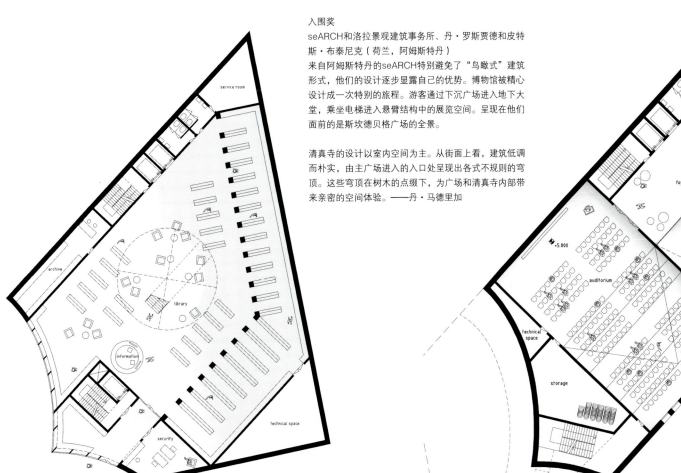

Finalist
Zaha Hadid Architects
London, UK

Design
Zaha Hadid with Patrik Schumacher

扎哈·哈迪德建筑事务所（英国，伦敦）

左图和下图
场地鸟瞰图
对页，上图
行人入口远景
对页，下图
立面图

Collaborators
Grant Associates Landcape
Architects, Buro Happold

LEFT AND BELOW
Aeriel view of site
OPPOSITE ABOVE
Pedestrian arrival perspective
OPPOSITE, BELOW
Elevations

The complex imagined by **Zaha Hadid Architects** takes the form of two seamless masses wrapping around the site perimeter, gradually growing in height from the museum to the mosque and culminating in a towering minaret. This layout creates an internal courtyard, an "intimate valley, a secluded garden for art, meditation and civic life" that also gives access to both the mosque and the museum.

The project, with its streamlined, curvilinear forms, bears the unmistakable aesthetic stamp of its renowned designer. In fact, Hadid even revisits one of her previous design concepts, the "urban carpet."

扎哈·哈迪德的设计由两个紧密连接的建筑，从低到高依次为：博物馆、清真寺和尖塔。这一布局形成了一个中央庭院，"一个私密的山谷、一座艺术、沉思和公众生活的花园"。从庭院同样可以进入清真寺和博物馆。项目的流线造型富有扎哈·哈迪德的显著特征。事实上，哈迪德甚至重新引用了她之前的设计理念——"城市地毯"。

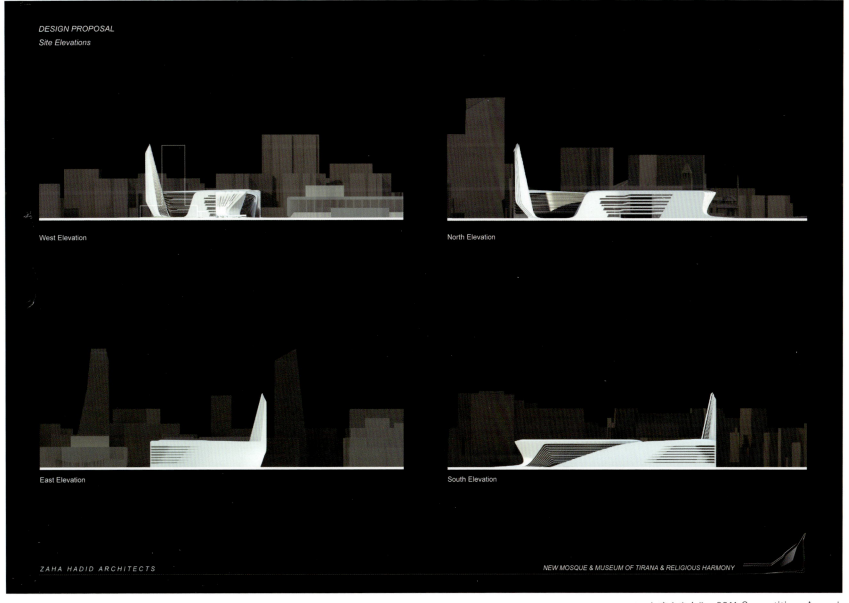

DESIGN PROPOSAL
Site Elevations

West Elevation

North Elevation

East Elevation

South Elevation

ZAHA HADID ARCHITECTS

NEW MOSQUE & MUSEUM OF TIRANA & RELIGIOUS HARMONY

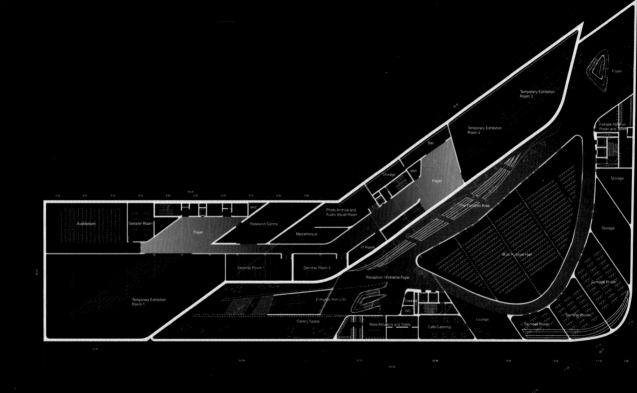

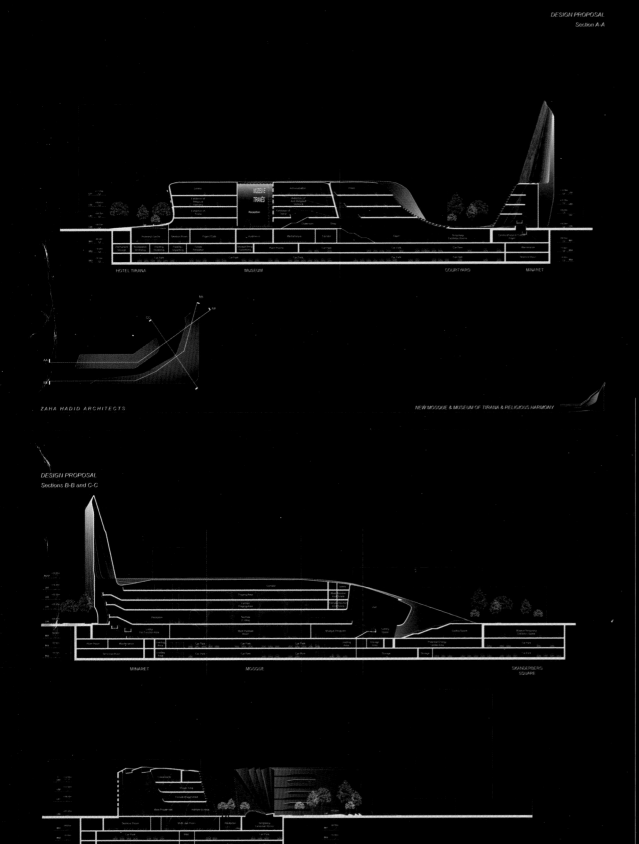

DESIGN PROPOSAL
Section A-A

ZAHA HADID ARCHITECTS

NEW MOSQUE & MUSEUM OF TIRANA & RELIGIOUS HARMONY

DESIGN PROPOSAL
Sections B-B and C-C

ZAHA HADID ARCHITECTS

NEW MOSQUE & MUSEUM OF TIRANA & RELIGIOUS HARMONY

入围奖
扎哈·哈迪德建筑事务所（英国，伦敦）

对页，上图
夜间鸟瞰图
对页，下图
平面图
本页
剖面图

Finalist
Zaha Hadid Architects
London, UK

Design
Zaha Hadid with Patrik Schumacher

Project Directors
Viviana Muscettola
Michele Pasca di Magliano
Loreto Flores, Effie Kuan

Project Team
Alvin Triestanto
Philipp Ostermaier
Hee Seung Lee
Gerry Cruz
Xia Chun
Ergin Birinci
Santiago Fernandez Achury
Rochana Chaugule
Soomeen Hahm
Chung Wang
Kanop Mangklapruk,
Luis Miguel Samanez
Alejandro Nieto

Collaborators
Grant Associates Landcape Architects
Buro Happold

OPPOSITE PAGE, ABOVE
Aeriel view at night
OPPOSITE BELOW
Floor plan
THIS PAGE
Sections

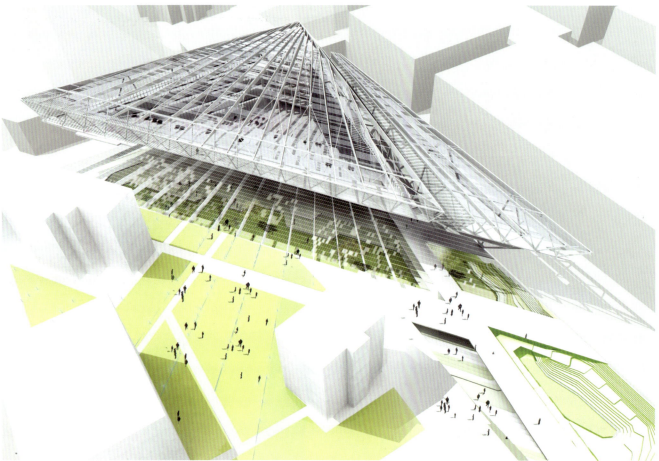

安德里亚斯·佩雷亚和NEXO（西班牙，马德里）

左图
入口鸟瞰图
下图
景观区域和花园
对页
植物园

Finalist: Andreas Perea Ortega & NEXO

Madrid, Spain

Design
Andreas Perea Ortega
NEXO: Ivan Carbajosa Gonzalez,
Lourdes Carretero Botran,
Manuel Leira Carmena

Collaborators
P.E.Z.+Adriana Giralt Landscape
Architects, Mecanismo (structral
engineers), Valladares (engineers),
Aurora Herrera (museum curator)

LEFT
Aerial view to entrance
BELOW
Landscaped area and garden
OPPOSITE PAGE
Botanical garden

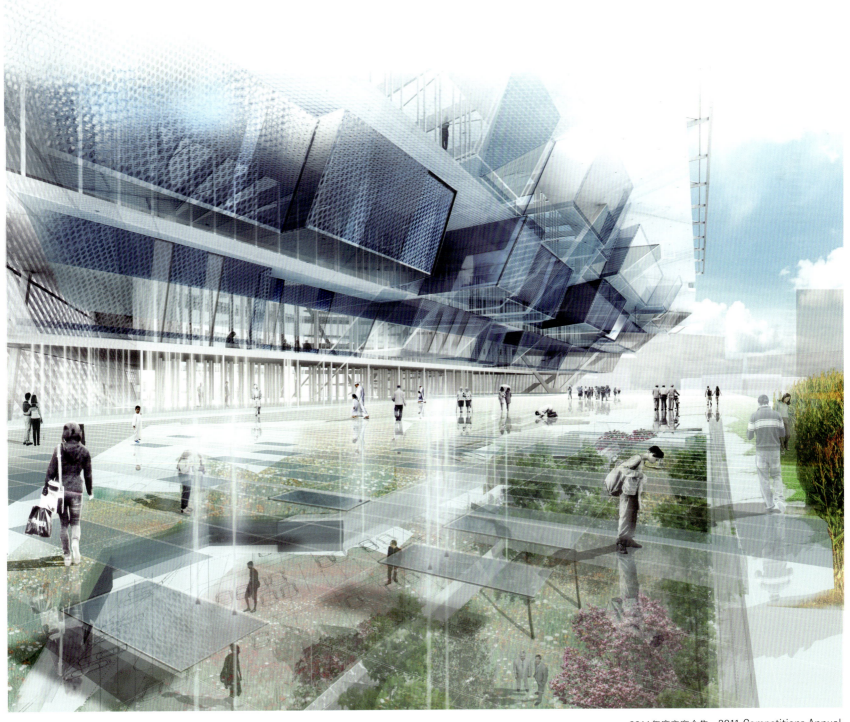

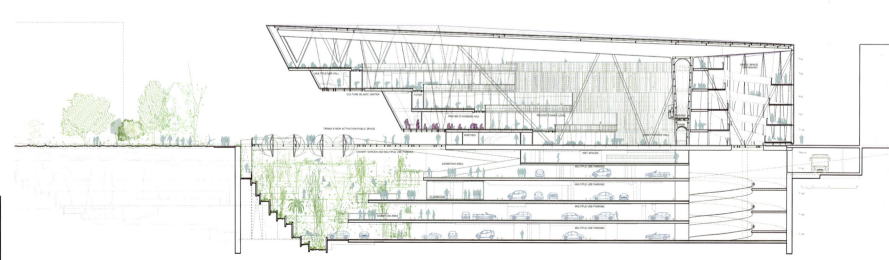

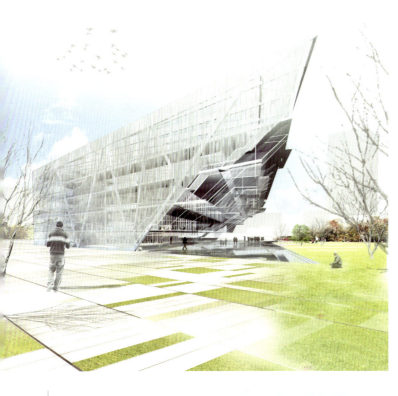

安德里亚斯·佩雷亚和NEXO（西班牙，马德里）
对页 清真寺室内和剖面
左图和下图 行人视角

Finalist: Andreas Perea Ortega & NEXO

Madrid, Spain
OPPOSITE PAGE
Mosque interior and section
LEFT AND BELOW
Pedestrian perspectives

Seemingly in opposition to Hadid's entry, the Ortega/NEXO team adopts a decidedly antistyle stance in an effort to "[transcend] those current architectural fashions that compete in a tiring parade of progressively more banal and puerile 'floral floats.'" The team posits that architecture "is not a problem of representing space. It is not a case of solving architectural problems with rhetorical solutions or formal or technological special effects doomed to outdate." With this position established, the team conceived a single volume cutting a strong diagonal across the site. In order to increase the volume of the public space a large portion of the complex is lifted from the ground to create a partially sheltered plaza that features a sunken, glass-covered garden. Inside, the mosque's vast prayer hall uses a system of tiers, oriented towards Mecca, that help gain the maximum amount of interior worship space with an efficiency that surpasses the schemes of the other finalists. Although the interior of the mosque is a refreshing, light-filled, albeit universal- leaning space, the highly structural aesthetic of the exterior gives few recognizable clues to its sacred function, and presents a rather steely coldness that is arguably ill matched to the objectives of the competition. -DM

安德里亚斯·佩雷亚和NEXO团队的设计与哈迪德的设计截然相反，采用了一个外观平和的设计，旨在"超越现有的建筑风格，在平庸幼稚的'植物'中脱颖而出"。该团队认为"建筑不是展现空间。无需用终将过时的装饰手法、造型或特殊工艺来解决建筑问题"。

设计团队采用单体造型进行设计。为了增加公共空间的面积，大部分楼体被从地面提起，形成了一个以下沉式玻璃花园为特色的广场。清真寺内部的朝圣大厅采用了层级系统，朝向麦加，拥有比其他围作品更宽敞的室内朝圣空间。尽管清真寺的内部是一个清新、明亮倾斜的空间，但是建筑外部并没有显著特征表明其功能价值，与其他竞赛作品相比，略显冰冷。——丹·马德里加

The permanently **opened door to everyone,** irrespective of race or religion, for all ages and through which not only the programmed activities are possible but which permits social and gregarious events to take place unexpectedly.

invention of a new public space

Architecture as a social and cultural engine

Andres Perea Ortega Architects | Nexo Architecture | Mecanismo Structural Builders | Valladares Consulting Engineers | Pilar Perea Artist | Andrea Giralt + P. E. Z. Landscape Architects

Expansion as an Art 艺术扩张
Daytona Museum of Arts and Sciences 代托纳艺术科学博物馆

现有的代托纳艺术科学博物馆入口（MOAS）
LEFT
View to entrance of present Daytona Museum of Arts and Sciences (MOAS)

Adding space to an existing museum to improve its functionality can be a daunting challenge. Confronted with such a scenario, the Daytona Museum of Arts and Sciences turned to a competition to arrive at an innovative solution to its expansion plans. Limited to architectural firms based in Florida, the competition was conducted in two stages — the first stage consisting of a short list based on expressions of interest, followed by a submission of designs by finalists.

The history of Daytona Museum of Arts and Sciences (MOAS) is similar to many museums, in that new wings were added to accommodate a larger collection. The level of the West Wing of the museum, located 30" below the main structure, can only be reached by a ramp, and is prone to flooding. To eliminate the need to move exhibits from this wing every time it is threatened by water, MOAS decided to demolish the existing wing and build a slightly larger structure to replace it at the same level as the rest of the museum complex. At the same time, they wanted to address the expansion of an entrance lobby, with the intention that it also be used for special events. The latter was considered to be a second phase if sufficient funding did not become immediately available. However, this latter phase of the program is certainly important to the image of MOAS, because it would provide it with a new sense of arrival for visitors.

As a multi-functional museum, MOAS is home to various types of activities and exhibits. In addition to a planetarium, its collection includes natural history, archeology, science, and art — Cuban, American, Afro-American, crafts and even a Coca

To administer the competition, MOAS engaged James Bannon, AIA, RIBA of DACORI Design and Construction, as a consultant. The subsequent RfQ limited to Florida firms resulted in three shortlisted firms as finalists:
• VOA , Orlando, Florida office
• HOK, Tampa, Florida office
• Architects Design Group, Winter Park, Florida

The jury was composed of 3 museum board members, the museum director, and an invited individual. The initial presentation by the teams was accompanied by comments from the jury, and the firms were then asked to refine their designs. When the final presentations took place, VOA was declared the winner, with ADG ranked second. -Ed

在原有博物馆的基础上进行功能扩建是一项重大的挑战。代托纳艺术科学博物馆试图在竞赛中寻求创新扩建方案。竞赛仅限佛里达州的建筑公司参加，分为两个阶段——第一阶段以提交意向书为基础，第二阶段则由入围者正式提交设计方案。

代托纳艺术科学博物馆的历史与其他博物馆相似，两侧翼楼的建造是为了收藏更多的藏品。博物馆的西翼低于主结构30英寸，只能通过坡道进入，并且容易在洪水时被淹没。为了避免每次水灾时都转移藏品，博物馆决定拆除现有的一楼，打造另一座较大的结构。同时，他们还希望扩建入口大厅，让它能够举办特殊活动。

作为一座多功能博物馆，代托纳艺术科学博物馆内能够举办各种类型的活动和展览。除了星象仪之外，它的藏品涉及自然历史、考古、科学和艺术，收藏范围从古巴、美国一直到非裔美洲。博物馆内甚至还有一个可口可乐展览。因此，博物馆的主要任务之一便是教育任务。

为了管理竞赛，代托纳博物馆邀请了詹姆斯·班诺（美国建筑师协会会员、英国皇家建筑师协会会员）作为顾问。最终，评审委员会选出了三支佛罗里达事务所入围：

VOA（佛罗里达州，奥兰多）

HOK（佛罗里达州，坦帕）

ADG（佛罗里达州，冬季公园）

评审委员会由三名博物馆董事会成员、博物馆馆长和受邀的独立个人组成。评审对团队的初步展示进行评论，随后各个团队进一步改善自身的设计。在最终展示中，VOA被宣布为获胜者，其次为ADG。——编者

优胜设计
（初始展示）
VOA（佛罗里达州，奥兰多）

左图 轴测图
下图，中间 行人视角
下图 剖面图

The Winning Design
(Initial presentation)

VOA
Orlando, Florida
Design Team
Jonathan Douglas, AIA
John Page, AIA
Daryl LeBlanc, Jay Jensen, Stephanie Moss,
Juan Gimeno, Rob Terry, George Mella,
Veronica Zurita, Fred Rambo

LEFT
Axonometric perspective
BELOW, MIDDLE
Pedestrian perspectives
BOTTOM
Section

The Winning Design
(Initial presentation)

VOA
Orlando, Florida

By moving the Planetarium from the interior of the MOAS to the entrance, the initial winning VOA proposal not only created an iconic arrival feature, but allowed for it to stay open for visitors when the rest of the museum is closed. When comparing VOA's original presentation with its final plan, the most notable change at the front entrance is the lower visibility given to the Planetarium due to its incorporation into the main structure, but still maintaining its own private entrance. According to Jonathan Douglas, VOA's team leader, the jury thought that VOA's initial presentation placed too much emphasis on architecture to the detriment of the art collection. Also, the interior "street" extending from the entrance to the new wing appears to be less grander in scale in the final scheme. -SC

优胜设计
（初始展示）
VOA（佛罗里达州，奥兰多）
VOA的优胜设计将星象仪从代托纳艺术科学博物馆的室内移到了门口，这不仅打造了一个地标性到达特征，还让它能够在博物馆关闭时向游客开放。在VOA初始展示和最终方案的对比中，最值得注意的是星象仪的视觉效果经过了削弱，被融入建筑主结构之中。根据VOA的领队乔纳森·道格拉斯所说，评审委员会觉得VOA的初始设计过多地强调了建筑而不是艺术收藏品。此外，从入口一直延伸到新翼楼的"室内街道"也没有初始设计那么宏伟。——斯坦利·科利尔

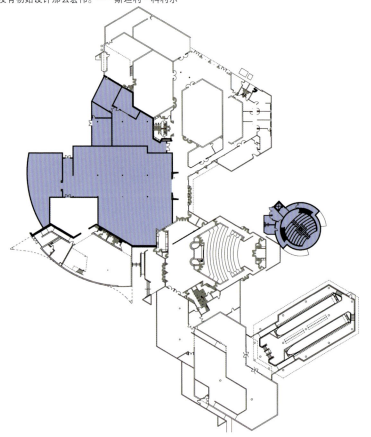

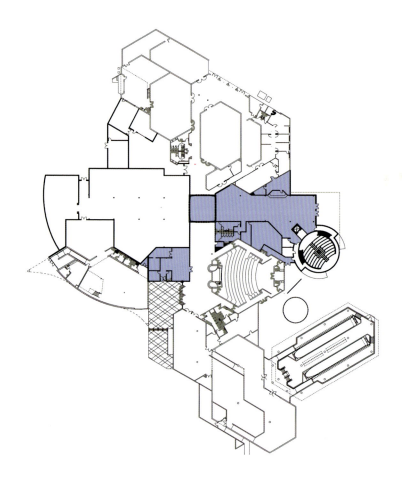

LEFT
Gallery
BELOW
Floor plan

OPPOSITE PAGE, ABOVE
Entry hall
OPPOSITE PAGE, BELOW LEFT TO RIGHT
Phases I & II construction

左图
画廊
下图
平面图

对页，上图
入口大厅
对页，下图由左至右
一期、二期工程

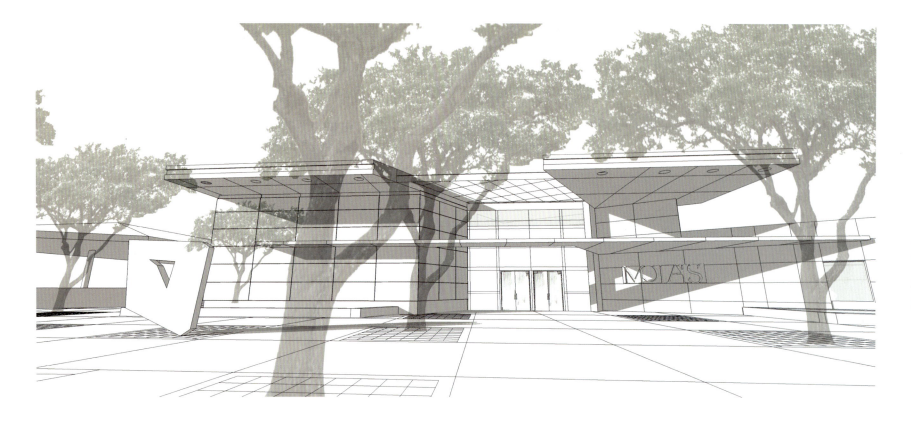

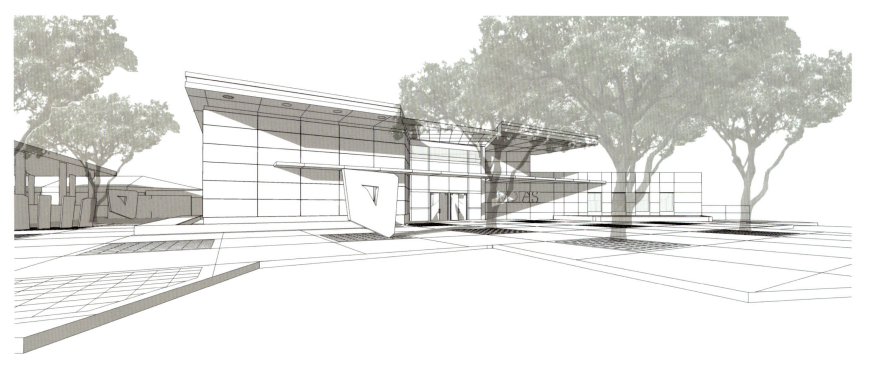

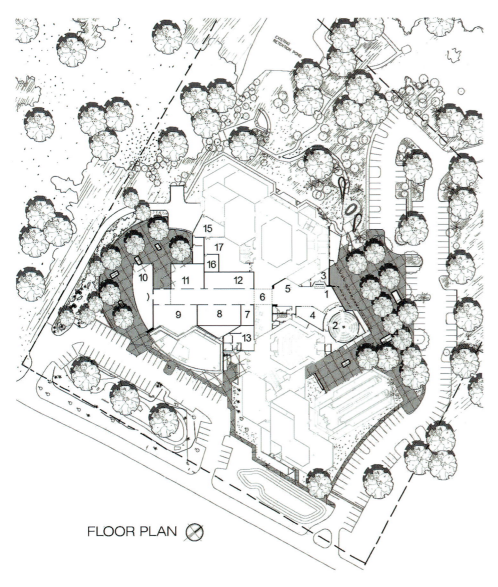

1   MAIN ENTRY

2   PLANETARIUM

3   ADMISSION & VOLUNTEER OFFICE

4   MUSEUM STORE

5   EXHIBIT SPACE

6   EXPLORATION CROSSROADS

7   MARZULLO GALLERY

8   CUBAN FOUNDATION MUSEUM

9   CHILDREN'S MUSEUM

10 THE CENTER FOR FLORIDA HISTORY

11 KARSHAN CENTER OF GRAPHIC ART

12 GILLESPY GALLERY

13 KITCHEN / PANTRY

14 CAFÉ

15 RECEIVING STUDIO

16 WORK ROOM

17 COLLECTION STORAGE

FLOOR PLAN

Further development proposal

VOA
Orlando, Florida

进一步开发方案

VOA（佛罗里达州，奥兰多）

LEFT
Floor plan
ABOVE, LEFT
Aerial view of model
ABOVE, RIGHT
Main hall

左图
平面图
左上
模型鸟瞰图
右上
主大厅

OPPOSITE PAGE, TOP AND MIDDLE
Views to entrance
OPPOSITE PAGE, BOTTOM
Section

对页，上图和中图
入口
对页，下图
剖面图

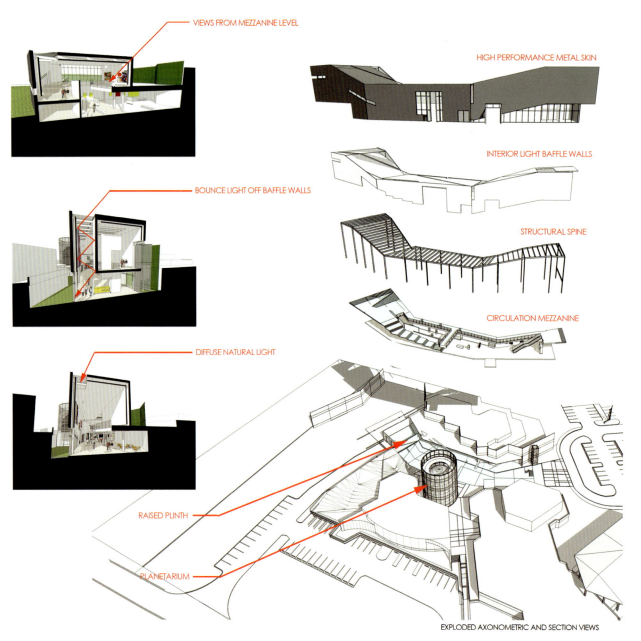

VIEWS FROM MEZZANINE LEVEL

BOUNCE LIGHT OFF BAFFLE WALLS

DIFFUSE NATURAL LIGHT

HIGH PERFORMANCE METAL SKIN

INTERIOR LIGHT BAFFLE WALLS

STRUCTURAL SPINE

CIRCULATION MEZZANINE

RAISED PLINTH

PLANETARIUM

EXPLODED AXONOMETRIC AND SECTION VIEWS

Second Place

Architects Design Group

Winter Park, Florida
Design Team
Keith Reeves V, FAIA – Principal in Charge
Susan Gantt, AIA – Project Manager
David Crabtree, Assoc. AIA – Project
Design Architect
Denis Vitoreli, - Intern Architect

Second Place **Architects Design Group** (ADG) from Winter Park concentrated the program along a central spine, being the only competitor to locate a major part of the program on a second level. This included a gallery for temporary exhibitions perched above the main entrance — part of the second phase expansion of a new entrance. Also, by including a second level, it also provided for access to a rooftop sculpture garden. The planetarium remained in the interior of the building, and, by using the two-tier plan, added space to the outside where the previous West Wing had been located.

二等奖

ADG（佛罗里达州，冬季公园）

来自冬季公园的ADG获得了二等奖，他们的设计以中央走道为中心，是唯一一个将主要项目设施设在二楼的方案。临时展览厅被设在主入口的上方，属于新入口扩建的二期工程。此外，二楼还通往一个雕塑花园。星象仪被保持在室内，通过双层设计，大大增加了原有的空间。
中央走道的设计让项目十分紧凑，汇集在较小的场地上面。

STUDY MODELS THAT EXPLORE FORM, FUNCTION, AND LIGHT

Second Place

Architects Design Group
Winter Park, Florida

FAR LEFT
models
LEFT, ABOVE
Plan
LEFT BELOW
Aerial view of model
BOTTOM, LEFT
first floor plan
BOTTOM RIGHT
Second floor plan

OPPOSITE PAGE, ABOVE
Axonometric and sections
OPPOSITE PAGE, BOTTOM LEFT
View to entrance
OPPOSITE, BOTTOM RIGHT
Main hall

二等奖

ADG（佛罗里达州，冬季公园）

左图三幅，模型
左上，规划图
左下，模型鸟瞰图
左侧底部，一层平面图
右侧底部，二层平面图

对页，上图，轴测图和剖面图
对页，左下，入口
对页，右下，主大厅

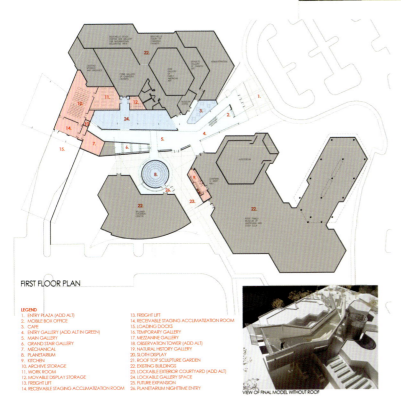

FIRST FLOOR PLAN

LEGEND
1. ENTRY PLAZA (ADD ALT)
2. MOBILE BOX OFFICE
3. CAFE
4. ENTRY GALLERY (ADD ALT IN GREEN)
5. MAIN GALLERY
6. GRAND STAIR GALLERY
7. MECHANICAL
8. PLANETARIUM
9. KITCHEN
10. ARCHIVE STORAGE
11. WORK ROOM
12. MOVABLE DISPLAY STORAGE
13. FREIGHT LIFT
14. RECEIVABLE STAGING ACCLIMATIZATION ROOM
15. LOADING DOCKS
16. TEMPORARY GALLERY
17. MEZZANINE GALLERY
18. OBSERVATION TOWER (ADD ALT)
19. NATURAL HISTORY GALLERY
20. SLOTH DISPLAY
21. ROOF TOP SCULPTURE GARDEN
22. EXISTING BUILDINGS
23. LOCKABLE EXTERIOR COURTYARD (ADD ALT)
24. LOCKABLE GALLERY SPACE
25. FUTURE EXPANSION
26. PLANETARIUM NIGHTTIME ENTRY

VIEW OF FINAL MODEL WITHOUT ROOF

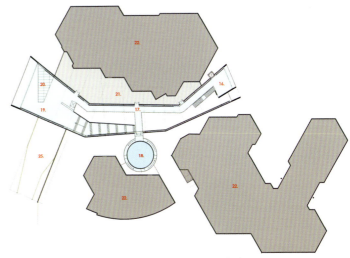

SECOND FLOOR PLAN

LEGEND
1. ENTRY PLAZA (ADD ALT)
2. MOBILE BOX OFFICE
3. CAFE
4. ENTRY GALLERY (ADD ALT IN GREEN)
5. MAIN GALLERY
6. GRAND STAIR GALLERY
7. MECHANICAL
8. PLANETARIUM
9. KITCHEN
10. ARCHIVE STORAGE
11. WORK ROOM
12. MOVABLE DISPLAY STORAGE
13. FREIGHT LIFT
14. RECEIVABLE STAGING ACCLIMATIZATION ROOM
15. LOADING DOCKS
16. TEMPORARY GALLERY
17. MEZZANINE GALLERY
18. OBSERVATION TOWER (ADD ALT)
19. NATURAL HISTORY GALLERY
20. SLOTH DISPLAY
21. ROOF TOP SCULPTURE GARDEN
22. EXISTING BUILDINGS
23. LOCKABLE EXTERIOR COURTYARD (ADD ALT)
24. LOCKABLE GALLERY SPACE
25. FUTURE EXPANSION
26. PLANETARIUM NIGHTTIME ENTRY

VIEW OF FINAL MODEL WITHOUT ROOF

HIGH PERFORMANCE METAL SKIN

INTERIOR LIGHT BAFFLE WALLS

STRUCTURAL SPINE

CIRCULATION MEZZANINE

DIFFUSE NATURAL LIGHT

BOUNCE LIGHT OFF BAFFLE WALLS

VIEW

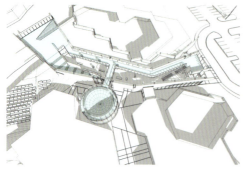

REBUILDING ANEW -
GREAT HALL OF THE ARTS AND SCIENCES

GREAT HALL = Traditionally, a great hall is the main room of a royal palace, nobleman's castle or a large manor house in the Middle Ages, and in the country houses of the 16th and early 17th centuries and many have been transformed into the museums of today. A typical great hall was a rectangular room between one and a half and three times as long as it was wide, and also higher than it was wide. It was entered through a screens passage at one end, and had windows on one of the long sides, often including a large bay window. There was often a minstrel's gallery above the screens passage. At the other end of the hall was the dais where the top table was situated. The lord's family's more private rooms lay beyond the dais end of the hall, and the kitchen, buttery and pantry were on the opposite side of the screens passage. Great Halls have served as large controlled collective spaces for entertainment, display, political free speech, and of course Art Collections for Palaces, Universities, private individuals, and Museums.

SITE AND ENVIRONMENT -
Nestled between the existing buildings and among the hydric swamp, the Great Hall weaves together a fractured complex of buildings. The ground level sits far above the water table protected from potential flooding. The building is designed with high performance glazing and exterior paneling that can resist wind, rain, and hurricane impact while having a super high efficiency. The hidden paths found within the hydric swamp served as an inspiration for the meandering form of the Great Hall as well as the control of diffuse natural light as found in the adjacent forest of filtered light. The sculptural metal skin is reflective for efficiency, but equally intended to be a sculptural element that creates a sense of arrival dignifies the Arts and Science.

GRAND ARRIVAL – INSPIRING THE SPIRIT OF THE ARTS AND SCIENCE
In part to inspire both young and the old, a grand arrival sets the stage for the events that will unfold and creates anticipation for what is to come. The large two story volume greets visitors as they approach from the dense hydric swamp. Once inside, the Grand Arrival space serves as the main entry, as well as, the terminus for the journey with in the Great Hall.

RAISED PLINTH

PLANETARIUM

DAYTONA BEACH MUSEUM OF ARTS AND SCIENCES ARCHITECTS DESIGN GROUP, INC. WINTER PARK • FORT MYERS

AXONOMETRIC VIEW

FIRST FLOO

Second Place

Architects Design Group

Winter Park, Florida

ABOVE AND RIGHT
Competition boards

二等奖

ADG（佛罗里达州，冬季公园）

上图和右图
竞赛展板

PRIMARY ENTRY THROUGH FORREST

MEANDERING ENTRY SEQUENCE

BRANDING SIGNAGE AND LIGHTING

MUSEUM ENTRY

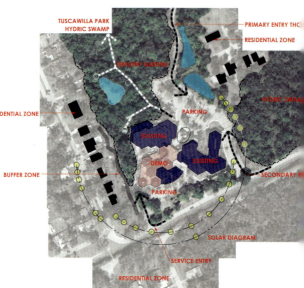

TUSCAWILLA PARK
HYDRIC SWAMP

PRIMARY ENTRY THC

RESIDENTIAL ZONE

SENSORY GARDEN

RESIDENTIAL ZONE

HYDRIC SWAM

PARKING

EXISTING

DEMO

EXISTING

BUFFER ZONE

SECONDARY E

PARKING

SOLAR DIAGRAM

SERVICE ENTRY

RESIDENTIAL ZONE

TE ENTRY SEQUENCE

SITE ANALYSIS

DAYTONA BEACH MUSEUM OF ARTS AND SCIENCES ARCHITECTS DESIGN GROUP, INC. WINTER PARK • FORT MYERS

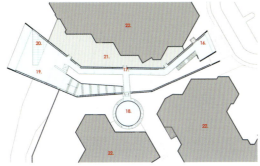

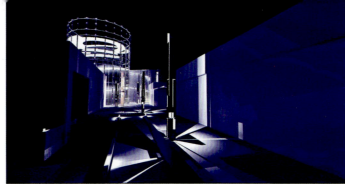
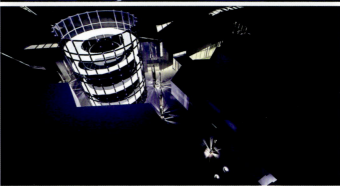

LEGEND
1. ENTRY PLAZA
2. MOBILE BOX OFFICE
3. CAFE
4. ENTRY GALLERY
5. MAIN GALLERY
6. GRAND STAIR GALLERY
7. MECHANICAL
8. PLANETARIUM
9. KITCHEN
10. ARCHIVE STORAGE
11. WORK ROOM
12. MOVABLE DISPLAY STORAGE
13. FRIEGHT LIFT
14. RECEIVABLE STAGING CLIMITAZATION ROOM
15. LOADING DOCKS
16. TEMPORARY GALLERY
17. MEZZANINE GALLERY
18. OBSERVATION TOWER
19. NATURAL HISTORY GALLERY
20. SLOTH DISPLAY
21. ROOF TOP SCULPTURE GARDEN
22. EXISTING BUILDINGS
23. EXTERIOR COURTYARD (WITH SLIDING SECURITY GATE)

CIRCULATION, PROCESSSION, AND WAYFINDING – THE GREAT HALL CONVERGES MANY DIVERGENT PATHS

The Great Hall serves as the main collection spaces for display, but more importantly, it creates a point of reference for the rest of the existing galleries within the Museum. The Grand Entry Gallery unfolds into the Main Gallery and likewise up the Grand Stair and into the Natural History and Sloth Display. A mezzanine level meanders back through the main gallery allowing and new perspective of art previously viewed. As visitors migrate to other areas within the Museum, they always have the Great Hall as a wayfinding device that unifies the Museum Complex.

FUNCTION AND FLEXIBILITY - THE CONTROL OF LIGHT AND INTERIOR ENVIRONMENT

The Great Hall concept serves many purposes ranging from permanent display, temporary display, entertainment, and other functions. The various gallery spaces within the Great Hall are all reconfigurable via movable walls, display cases and there is gracious wall space for wall mounting various exhibits of both small and large scale. Controlled natural light is allowed into some areas via high performance translucent glazing systems and this is further complement by interior lighting and interior environment control systems.

MOTAS
MUSEUM OF ARTS & SCIENCES

SECOND FLOOR PLAN

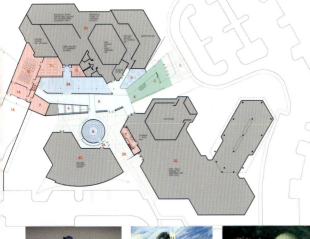

LEGEND
1. ENTRY PLAZA (ADD ALT)
2. MOBILE BOX OFFICE
3. CAFE
4. ENTRY GALLERY (ADD ALT IN GREEN)
5. MAIN GALLERY
6. GRAND STAIR GALLERY
7. MECHANICAL
8. PLANETARIUM
9. KITCHEN
10. ARCHIVE STORAGE
11. WORK ROOM
12. MOVABLE DISPLAY STORAGE
13. FRIEGHT LIFT
14. RECEIVABLE STAGING ACCLIMATEATION ROOM
15. LOADING DOCKS
16. TEMPORARY GALLERY
17. MEZZANINE GALLERY
18. OBSERVATION TOWER (ADD ALT)
19. NATURAL HISTORY GALLERY
20. SLOTH DISPLAY
21. ROOF TOP SCULPTURE GARDEN
22. EXISTING BUILDINGS
23. LOCKABLE EXTERIOR COURTYARD (ADD ALT)
24. LOCKABLE GALLERY SPACE
25. FUTURE EXPANSION
26. PLANETARIUM NIGHTTIME ENTRY

PLAN

SECOND FLOOR PLAN

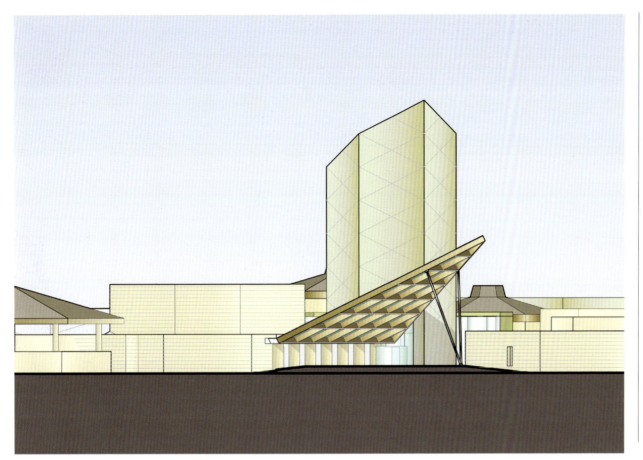

Third Place

HOK
Tampa, Florida

HOK's plan was probably the most straightforward in that it made the fewest internal changes, designating the former West Wing area as a large exhibition space. The idea here was flexibility, as the exhibition space could be configured to accommodate either one large, or several smaller exhibits. The iconic architectural feature of the HOK plan was a Sun Tower at the entrance, intended to raise the museum's visibility in the neighborhood and from the distant road.

三等奖

HOK（佛罗里达州，坦帕市）

HOK的设计或许是最直观的。它几乎对室内没做改动，将西翼区域设置成为一个大型展览空间。设计概念十分灵活，展览空间可以整合为一个大型空间，或分割成若干个小型展览空间。HOK设计的标志性特征是入口处的太阳塔。这一设计提升了博物馆的视觉效果。

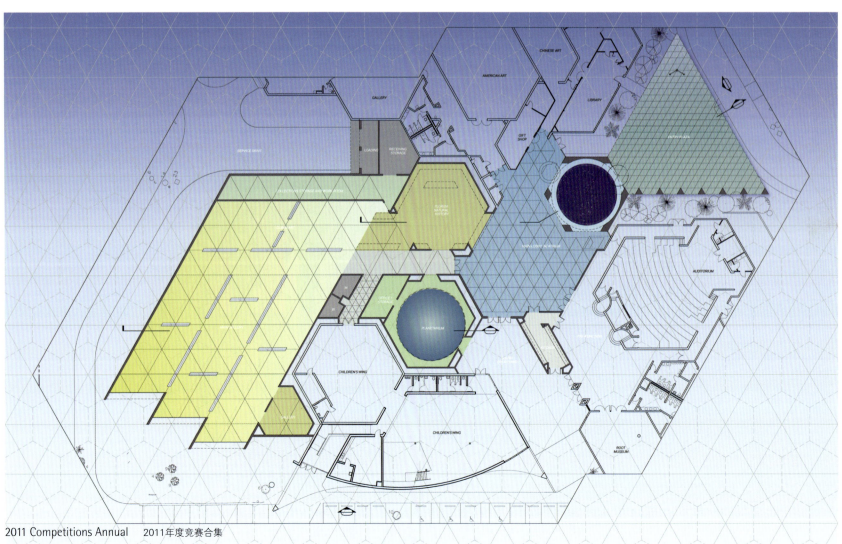

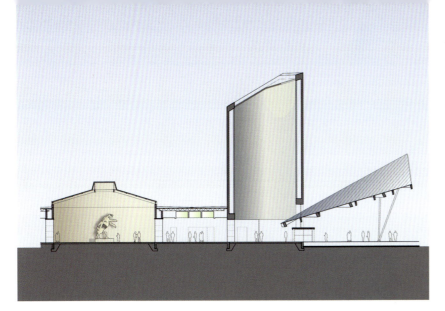

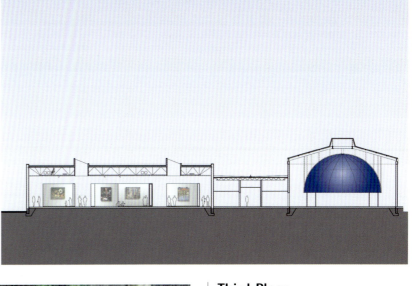

Third Place

HOK

Tampa, Florida
Design Team
Yann Weymouth, AIA, LEED AP, Senior Vice President, Design Principal
Michael Harris, AIA, LEED AP, Associate, Project Designer

ABOVE
Sections
LEFT
Site plan
OPPOSITE PAGE, AGOVE
Lobby view
OPPOSITE PAGE, BELOW
Floor plan

三等奖

HOK（佛罗里达州，塔帕）

上图
剖面图
左图
总规划图
对页，上图
大堂
对页，下图
平面图

Museum as Sustainability Model 作为可持续模型的博物馆

1. BUILDING AS LANDSCAP
與景觀合而為一的建筑
A topographic phenomen
地形的變

The Taipei City Art Museum Competition 台北市立艺术馆竞赛

To arrive at a design for a new art museum in Taipei, the organizers decided to allow the participants more flexibility than usual in devising their planning concepts for the new institution. According to the design brief, "the planning and design guidelines in this program are for reference only. The designer must propose...new possibilities for modern art museums, define the exhibition method, and propose new space requirements, then proceed (in) the planning and design based on the new required spaces and design guidelines."

The Jury Process

As for the finalists, all three had quite different approaches, albeit leaning heavily on sustainability as a major tenet. The choices varied from the linear and vertical to historicism: wrapping, burying and rooftop park were the schemes that caught the jury's eye. The finalists were:

* **Peter Boronski / Jean-Loup Baldacci, New Zealand/France**
* **Kengo Kuma / Kengo Kuma & Associates, Japan**
* **Federico Soriano Pelaez, Spain**

Five honorable mentions who did not take part in the second stage were:

* **James Law Cybertecture International Holding** (Hong Kong)
* **Sung Goo Yang with Oscar Kang** (USA)
* **Ian Yan-Wen Shao with JR-Gang Chi, Ar-Ch Studio** (Taiwan)
* **Jafar Bazzaz** (Iran)
* **Ysutaka Oonari Masamichi Kawakami, 101 Design** (Japan)

三件入围作品虽然都以可持续性为主题，侧重方向却截然不同，从线型、垂直到历史相对论：包裹、掩藏和屋顶公园分别吸引了评审的眼球。三个入围者分别为：
· 彼得·博隆斯基/让-卢·巴尔达西（新西兰/法国）
· 隈研吾/隈研吾建筑事务所（日本）
· 费德里克·索里亚诺·佩雷斯（西班牙）
五个未参与第二阶段竞赛的荣誉奖获得者分别为：
· 科建国际有限公司（香港）
· 成谷阳和奥斯卡·康（美国）
· 邵彦文和漆志刚（Ar-Ch建筑工作室）（台湾）
· 贾法尔·巴扎斯（伊朗）
· 欧纳利·雅隆和川之上雅道（101设计）（日本）

一等奖
彼得·博隆斯基/让-卢·巴尔达西（新西兰/法国）
设计根据地势起伏叠垒，与外国工作室、乃至扎哈·哈迪德的设计有异曲同工之妙。这件作品展示了设计与周边公园式景观的强烈联系。根据设计师所说："这不是一个简单的美术馆项目，它层叠的造型形成了压缩或扩张的室内外空间，是一个类似'城市田地'的空间。它更像是一个'意识流'作品，成为了一个标志性元素。"

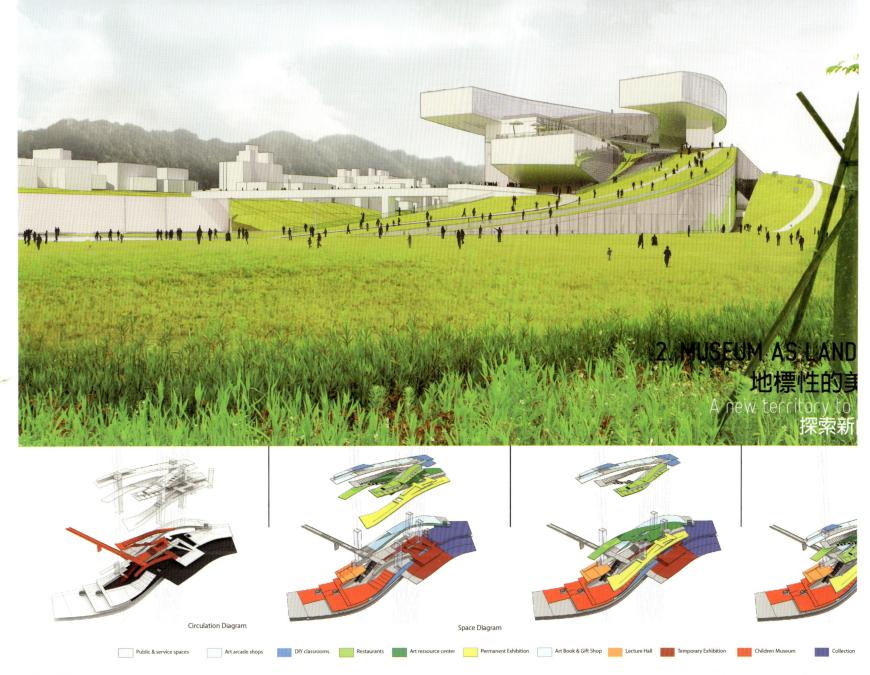

Circulation Diagram

Space Diagram

Public & service spaces   Art arcade shops   DIY classrooms   Restaurants   Art ressource center   Permanent Exhibition   Art Book & Gift Shop   Lecture Hall   Temporary Exhibition   Children Museum   Collection

First Place

Peter Boronski / Jean-Loup Baldacci

New Zealand/France

Composed of a layered, undulating configuration tied together by the topography, one wonders if the jury had the firm, Foreign Office, or even Zaha Hadid in mind when it selected this design for the final round. Despite such superficial comparisons, this entry suggests a strong connection to a park-like setting, a place to be visited on weekends and holidays with the entire family. As an extension of the park, "It is completely accessible for people to walk and even ride bicycles all over." According to the authors, "This is not a museum as singular object but rather a field of overlapping volumes—surfaces that form compressing and expanding interior and exterior spaces, a quasi-urban "field" to wonder on. It is more a "stream of consciousness" to dive into than a building as signature object."

The thought driving the configuration and circulation was a rather

这种造型和配置的驱动力来自解决复杂方案的简单想法："正如影响台湾民主政策的历史、政治和文化因素一样，美术馆以不规则的流动元素形式展开。像下方的河流一样弯曲、交叉。这些元素在滑动、膨胀和浮现的过程中容纳各种功能设施。它们的隔断与众不同，中间设有空隙和空间。它们仿佛在不断运动之中，位置不断变化，给人以影院式的体验。美术馆内没有静止。"

simple idea resulting in a complex solution: "Like the various historical, political and cultural influences that have been converging and working to sharp the moderate democracy of Taiwan, this museum sweeps up out of the ground in a dis-array of fluid elements. Curving and crossing like the waters of converging Yingge & Dahan rivers below. There elements are containers that struggle to contain, as they themselves slip, bulge and emerge. There are compartments but they are not regular, and there are volumes and voids. Their understanding requires movements, a changing of positions, like a cinematic experiences. Stasis is not comfortable in this house."

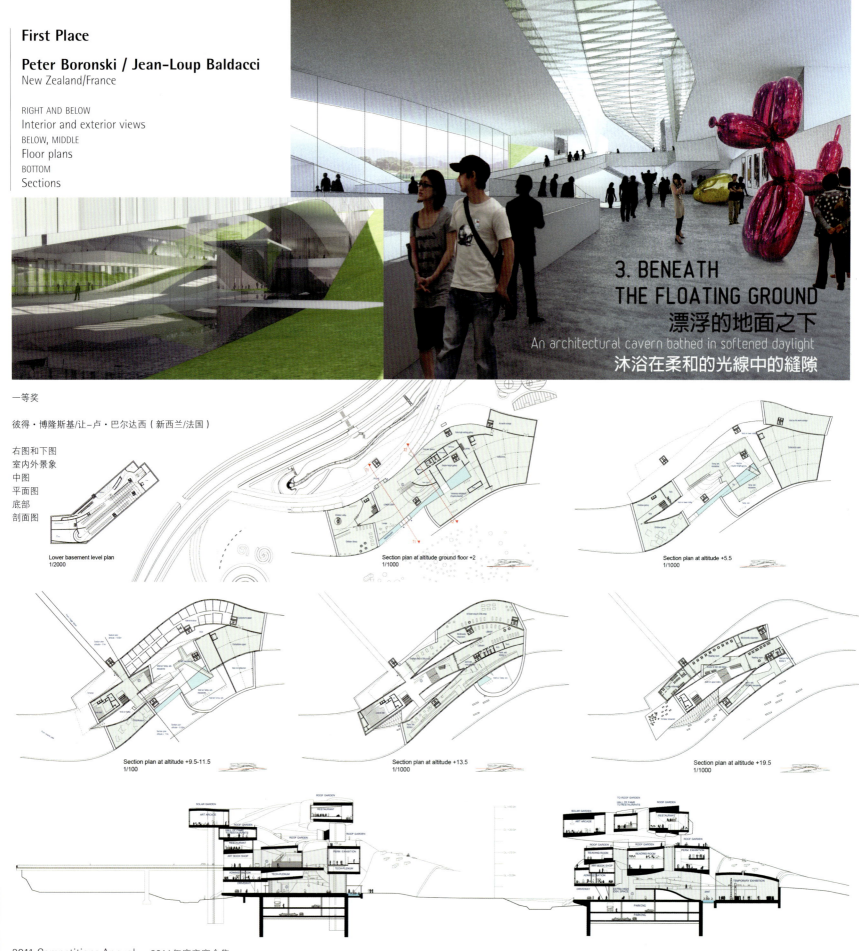

First Place

Peter Boronski / Jean-Loup Baldacci
New Zealand/France

RIGHT AND BELOW
Interior and exterior views
BELOW, MIDDLE
Floor plans
BOTTOM
Sections

3. BENEATH
THE FLOATING GROUND
漂浮的地面之下
An architectural cavern bathed in softened daylight
沐浴在柔和的光線中的縫隙

一等奖

彼得·博隆斯基/让-卢·巴尔达西（新西兰/法国）

右图和下图
室内外景象
中图
平面图
底部
剖面图

Lower basement level plan
1/2000

Section plan at altitude ground floor +2
1/1000

Section plan at altitude +5.5
1/1000

Section plan at altitude +9.5-11.5
1/100

Section plan at altitude +13.5
1/1000

Section plan at altitude +19.5
1/1000

8. MoA NEW TAIPEI CITY
新北市立美術館
A field of dreams
一個夢想的園地

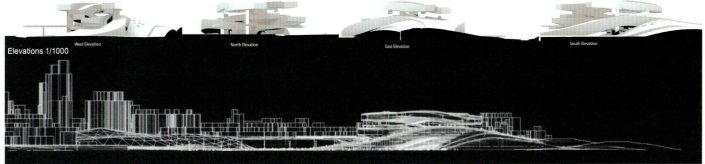

Elevations 1/1000

West Elevation   North Elevation   East Elevation   South Elevation

LEFT
Night view from city
Elevations
BELOW FROM LEFT CLOKWISE
Birdseye view
Museum interior
Approach at night
View to lobby

5. ARTISTIC VILLAGE
藝術村莊
Embracing diversity
結合多元的活動

Second Place

Kengo Kuma / Kengo Kuma & Associates
Tokyo / Paris

二等奖

隈研吾/隈研吾建筑事务所（东京/巴黎）

Here the similarity to Frank Gehry can hardly be avoided, apart from the material nature of the wrapping enclosing the structure. As icon, this entry represents a truly strong urban statement. Organized around a "green cell" system consisting of diverse materials and aggregated in different functions to match the programmatic requirements. The double skin is to play a principal role in the sustainability process providing an interesting "in-between space." Here again, the notion of erasing the boundaries between private and public space is strong, the idea being to "provide the citizens with an informal experience of contemporary art."

Although sustainability is high on the list of priorities here; lurking in the background will always be the question of budget viability. This may have been one of the reasons this rather daring proposal was on the outside, looking in.

除了结构外部采用天然材料之外，该设计与弗兰克·盖里设计的相似性难以忽略。作为一个地标，这件作品呈现了强烈的城市表达。项目围绕一个"绿色细胞"系统展开，将各种各样不同的材料聚合在一起来满足项目要求。设计同样注重消除私人和公共空间的界限，表达了"为市民提供通俗的现代艺术体验"的概念。

尽管设计极其注重可持续性，将美术馆融入背景仍然需要庞大的预算。这也许就是这个大胆的方案没有获得优胜的原因。

Green Cell

New tower museum uses Green Cell as a symbol for Yingge and New Taipei City

The cells consist of diverse materials, aggrupated in different functions, in order to adapt to different programmatic needs.

A new technology based on a double skin is proposed for the museum. It serves as strucure but also plays a key role in environmental sustainability.

It provides a chance to regenerate and redefine the urbanscape.

New Taipei City Museum of Art Conceptual Design　　International Design Competition　　13th October 2011　　KKAA - Kengo Kuma & Associates　　1

Site Plan

CELLS
Ground Floor Texture Cell　　Furniture Cell
wood　grass　stone　　table and chair
0　10　30　60　　100(m²)

Green Cell towards city
Green Cell re-generate Yingge

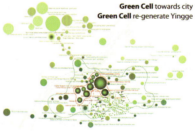

Isometric view

Contemporary Art Museum

Administration

Children's Art Museum　　Lecture Hall

Education Center

Green Cell's roof

Green Cell in Yingge

The concept for program distribution emerges from the notion of absorbing the private and public space, attempting to merge and activate both city, park and museum, providing the citizens with an informal experience of contemporary art.

In opposition to the verticality of the tower, its counterpart - the Main Hall - acts as communications hub. A lively space that serves all the possible emerging purposes - leisure, meeting point, exhibition, art workshop, parties... A place for people to use however they desire.

The interior space continues vertically through the interlacing of exhibition, public and commercial space.

A truly social museum, an whirlpool of art, artists and public.

Fertile ground for creation, invitation, and sprouting of a new culture.

shops　　bus stop
pavilion
gallery

Green Cell / extension to city
We believe that this museum will not end in the mere exhibition of art. The CELL serve as a strategy to add more significance to the urban tissue, emerging as a new symbol. A pavilion in the park, shops, urban furniture and other public programs can be embraced by the pattern.

Green cell / a center of Yingge
The main hall connects not only the museums but manages all the fluxes of the territory: train station, cable car, riverbank path, city and museum merge in this void, energizing the urban space.

Green Cell / Versatility
Steel mesh Green Cells besides structural elements, they also serve as a framework for the skin which accomodates greenery, soil, LED screens, ETFE and photovoltaic pannels.

ventilation
ETFE　　LED screen
photovoltaic cell　　green panel

Green Cell / in-between space
The use of a double layer skin creates an in-between space which increases the flexibility for exhibition, and addresses the specificities of the Taiwanese climate.

Outer Skin
Inner Skin
Air Condition
Extension

Green Cell in Yingge

View From Front Path

New Taipei City Museum of Art Conceptual Design　　International Design Competition　　13th October 2011　　KKAA - Kengo Kuma & Associates　　2

KKAA - Kengo Kuma & Associates　　3

1F Plan (±0.00)

from the station

Bridge

to the Main Hall

to the Museum

Security
Lobby
Shop
Ticketing
Backyard
Main Hall
Info
Big Stair
Shop
to the Lecture Hall

Library
Waiting area
to parking
B3 parking
Lobby
Foyer
Dry Area
Lounge
Lecture Hall

to the Main Hall

Tour Bus Parking
(10 cars)

Cooling Tower

Cable Car

CELLS
Ground Floor Texture Cell   Furniture Cell
wood   grass   stone   table and chair

19F Plan (100.00) 19.5F Plan (103.50) 20F Plan (108.00) 20.5F Plan (111.50)

16F Plan (83.50) 17F Plan (88.00) 18F Plan (92.50) 18.5F Plan (96.00)

12F Plan (62.50) 13F Plan (68.50) 14F Plan (74.50) 15F Plan (79.00)

8F Plan (39.50) 9F Plan (45.50) 10F Plan (50.50) 11F Plan (56.50)

4F Plan (16.50) 5F Plan (22.50) 6F Plan (27.50) 7F Plan (33.50)

Car & Bus Stop Area

New Taipei City Museum of Art Conceptual Design   International Design Competition   13th October 2011

Big Hall

New Taipei City Museum of Art Conceptual Design   International Design Competition

HYBRID VENTILATION COOLING SYSTEM

Damper   Exhaust fan

AHU

Natural Ventilation   Mechanical Ventilation   Slightly Cooling Ventilation

AIR CONTROL SYSTEM

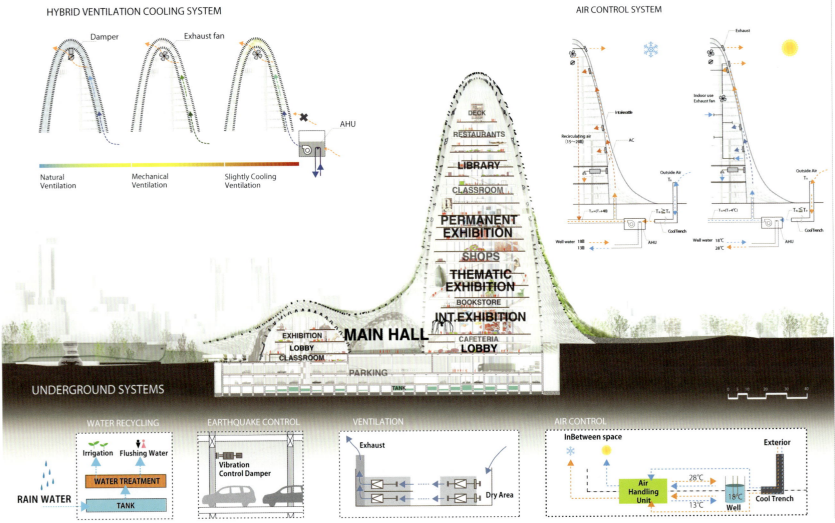

DECK

RESTAURANTS

LIBRARY

CLASSROOM

PERMANENT EXHIBITION

SHOPS

THEMATIC EXHIBITION

BOOKSTORE

INT. EXHIBITION

EXHIBITION LOBBY CLASSROOM   MAIN HALL   CAFETERIA LOBBY

PARKING   TANK

UNDERGROUND SYSTEMS

WATER RECYCLING

Irrigation   Flushing Water

WATER TREATMENT

RAIN WATER   TANK

EARTHQUAKE CONTROL

Vibration Control Damper

VENTILATION

Exhaust

Dry Area

AIR CONTROL

InBetween space   Exterior

Air Handling Unit   28°C

13°C   18°C   Well   Cool Trench

Second Place

Kengo Kuma / Kengo Kuma & Associates

Tokyo / Paris

二等奖

隈研吾/隈研吾建筑事务所（东京/巴黎）

KKAA - Kengo Kuma & Associates

3

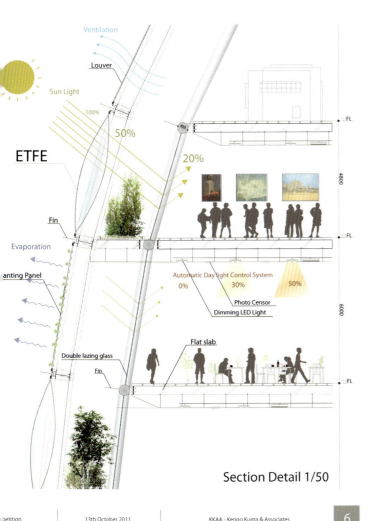

ETFE

Section Detail 1/50

In-Between Space

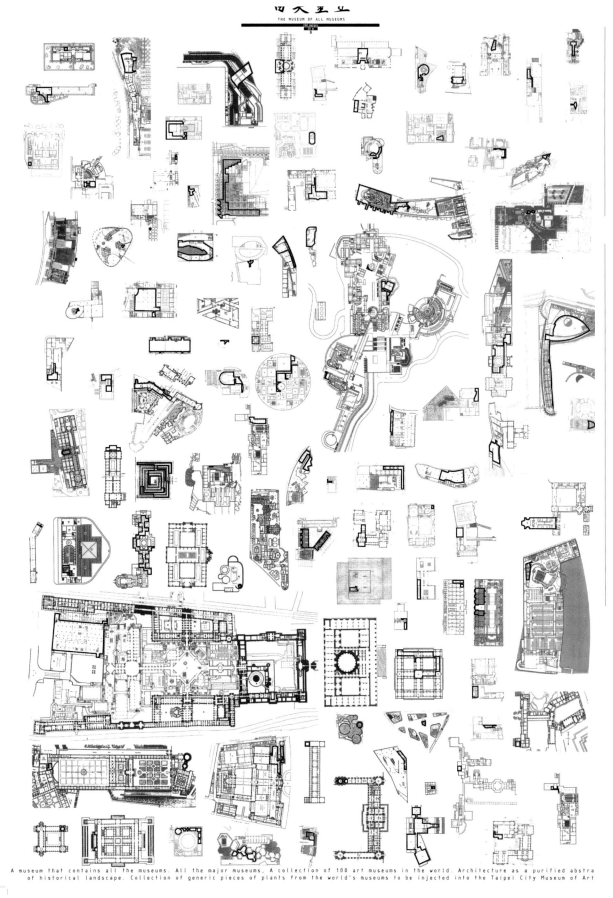

THE MUSEUM OF ALL MUSEUMS

A museum that contains all the museums. All the major museums, A collection of 100 art museums in the world. Architecture as a purified abstra of historical landscape. Collection of generic pieces of plants from the world's museums to be injected into the Taipei City Museum of Art

Third Place

Federico Soriano Pelaez
Madrid, Spain

LEFT
Collage of existing museum plans
OPPOSITE, ABOVE
View of above-grade structures toward the museum entrance
OPPOSITE, BELOW
Sections

The **Federico Soriano Pelaez entry** is based in part on the idea that bits and pieces of past museum plans can be adapted to a new museum. Fitting some of these isolated parts into a cohesive whole may be a fascinating exercise, but it does beg the question: Who can possibly view these plans other than a construction project manager. Since most of the museum is buried underground, one might imagine that this is something Peter Eisenman might cook up, whereby the organizational idea is hidden from the view of the average pedestrian.

Another problem is the non-iconic nature of the proposal. Considering that the initial impression of the visitor upon arrival (forget the signage) might be that of a campus—or even some modern version of an Italian cemetery, what a surprise to find that everything is located in one continuous underground structure.

There can be no doubt that this building would have attained a high LEED rating by virtue of being primarily below grade, leading to a very low energy consumption. -SC

三等奖
费德里克·索里亚诺·佩雷斯（西班牙，马德里）

左图
现有美术馆方案的拼贴画
对页，上图
朝向博物馆入口的上层结构
对页，下图
剖面图

费德里克·索里亚诺·佩雷斯的设计以各式各样的美术馆方案的拼接为接触。将这些独立的部分拼贴成一个整体将是一次有趣经历，但是也产生了一个问题：除了工程经理之外，谁还能看过这么多的方案。由于大多数博物馆都已经被掩埋在尘土之下，人们可能会想象这只是彼得·埃森曼编造出来的，反正行人们不会发现地下的组织结构。
另一个问题是该设计不具有标志性。来访者的第一印象会认为这是一个校园或是一座意大利现代公墓，很难发现所有的设施都在地下结构中。
毫无疑问，这座建筑能够获得绿色建筑最高标准认证。由于所有设施几乎都在地下，美术馆的能源消耗非常低。——斯坦利·科利尔

The cross-sections show the two exhibition levels resting on the solid foundation of the parking lots and the building's mechanised systems. This liberated the roof and the emerging volumes making them landscape elements over the garden.

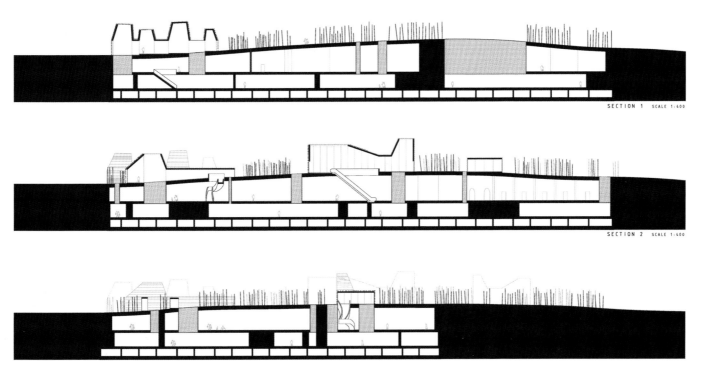

SECTION 1 SCALE 1:400

SECTION 2 SCALE 1:400

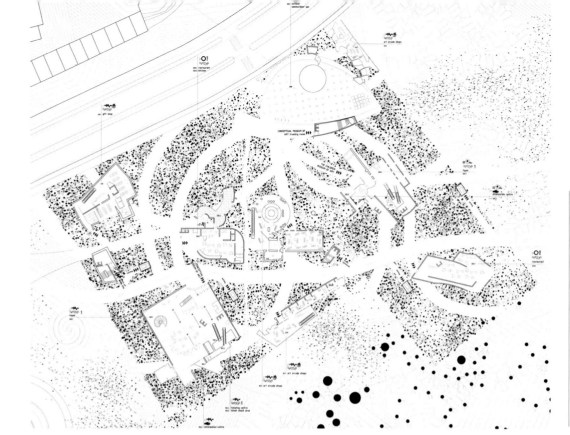

三等奖

费德里克·索里亚诺·佩雷斯（西班牙，马德里）

左图，地面层
左下，地下一层
右下，地下二层
对页，上图
美术馆鸟瞰图
对页，下图
立面图

Third Place

Federico Soriano Pelaez
Madrid, Spain

LEFT
Grade level plan
BELOW, LEFT
1st level below grade
BELOW, RIGHT
2nd level below grade
OPPOSITE, ABOVE
Birdseye view of museum
OPPOSITE, BELOW
Elevations

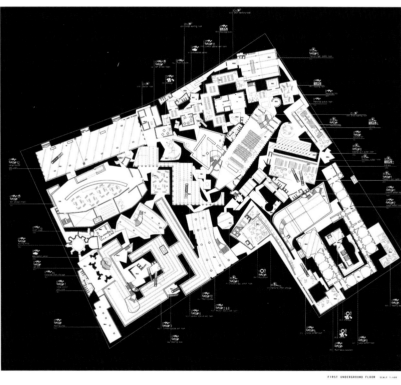

FIRST UNDERGROUND FLOOR   SCALE 1:400

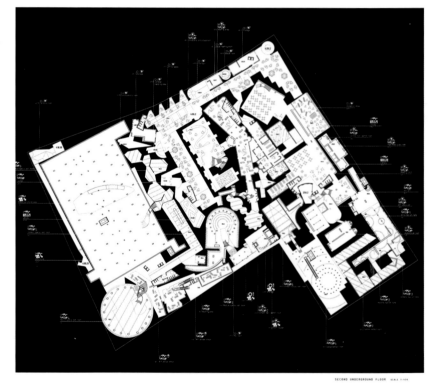

SECOND UNDERGROUND FLOOR   SCALE 1:400

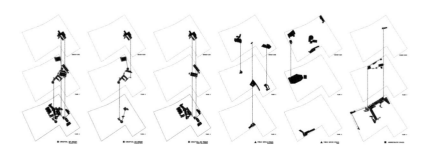

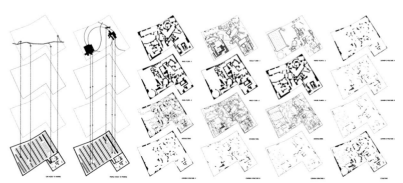

The wind sways the flexible stems. The public moves back and forth as another branch. In the background a building emerges. We approach it enjoying the scents. In the distance, the buildings rise above the sea of greenery and are visible from the city.

WEST ELEVATION SCALE 1:400

NORTH ELEVATION SCALE 1:400

EAST ELEVATION SCALE 1:400

SOUTH ELEVATION SCALE 1:400

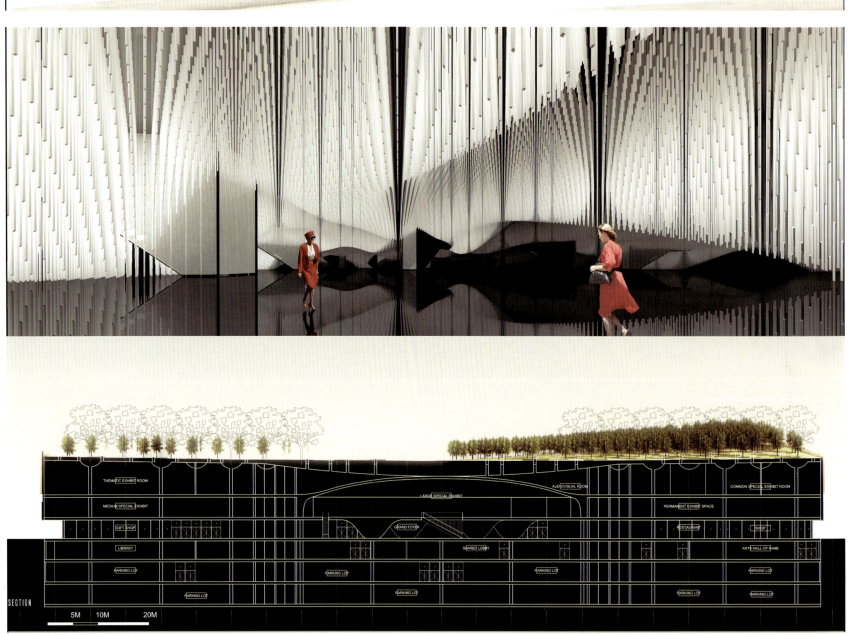

荣誉奖1

成谷阳和奥斯卡·康（美国）

本页和对页
图片来自于竞赛展示板

Honorable Mention 1

Sung Goo Yang
with Oscar Kang
USA

THIS PAGE AND OPPOSITE
Images from competition
boards

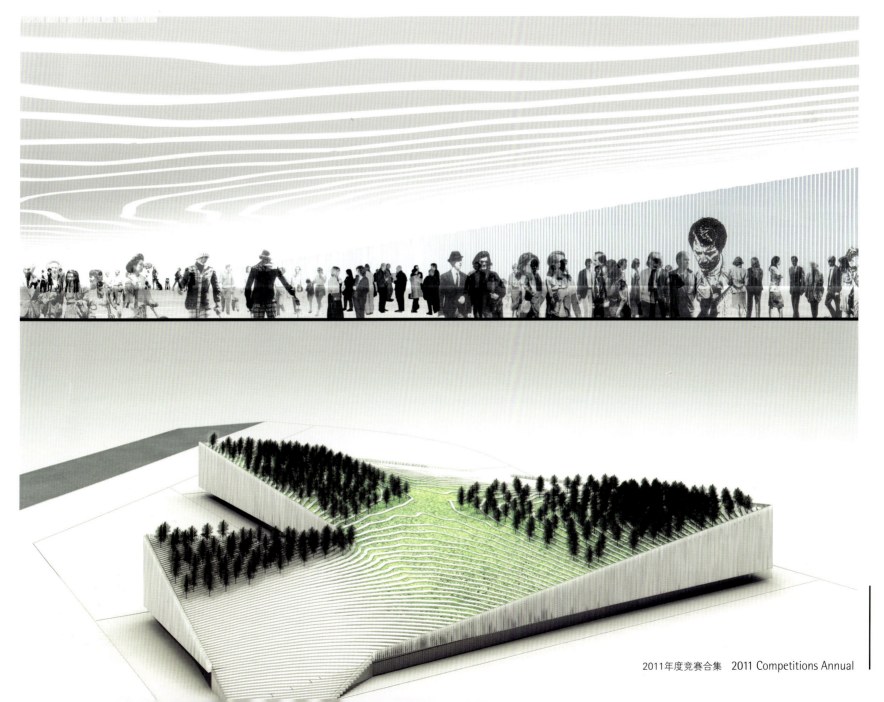

Moving Galleries
Dynamic Museum
New Taipei City Museum of Art

Dynamic Museum consists of moving galleries.
Some of these galleries move up/down in vertical direction.
Others move left/right horizontally.
Moving galleries create dynamic space that changes throughout time.
Visitors are able to experience different views of interior and exterior spaces.
Museum building reflects its unique cinematic dramatic character.

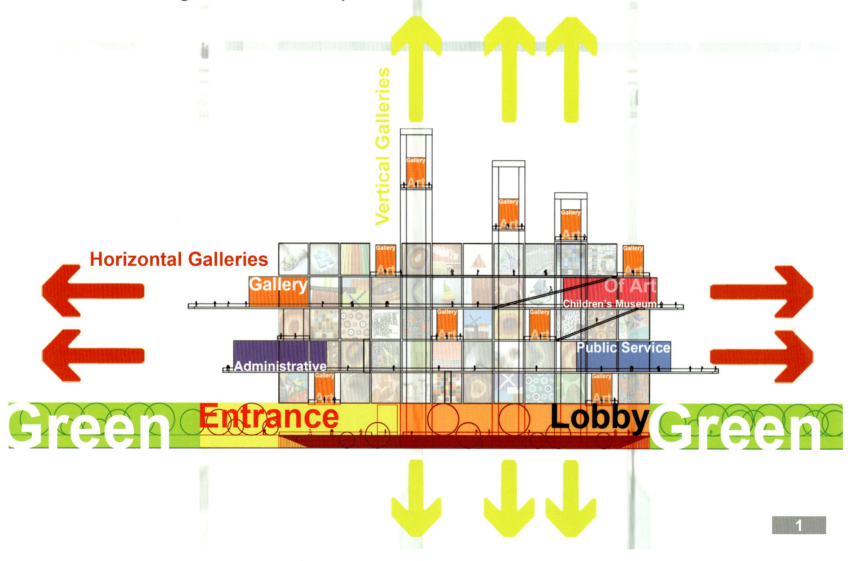

1

荣誉奖2

贾法尔·巴扎斯（伊朗）

本页和对页
图片来自于竞赛展示板

Honorable Mention 2

Jafar Bazzaz
Iran

THIS PAGE AND OPPOSITE
Images from competition boards

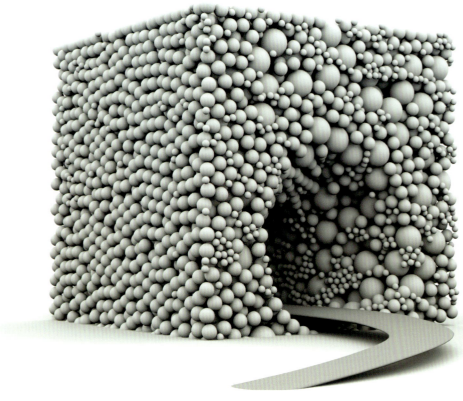
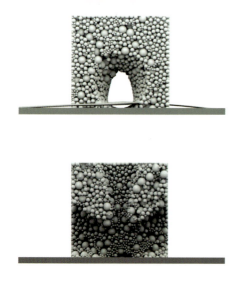
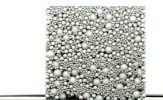

DESIGN CONCEPT

States of matter are the distinct forms that different phases of matter take on. Solid is the state in which matter maintains a fixed volume and shape; liquid is the state in which matter maintains a fixed volume but adapts to the shape of its container; and gas is the state in which matter expands to occupy whatever volume is available.

SOLID　　　**LIQUID**　　　**GAS**

FORM

A symmetrical three-dimensional shape, either solid or hollow, contained by six equal squares.

LIQUID

Form can be adjusted to the object if nothing is rigid. It is formless like water. It becomes whatever it is contained in. It is flexible, fluid and promotes motion.

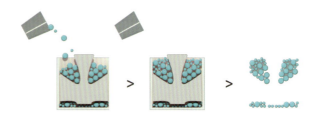

CUBE

A symmetrical three-dimensional shape, either solid or hollow, contained by six equal squares. ter maintains a fixed volume but adapts to the shape of its container; and gas is the state in which matter expands to occupy whatever volume is available.

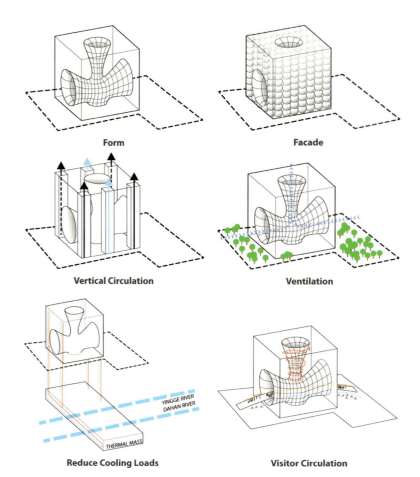

Form　　　**Facade**

Vertical Circulation　　　**Ventilation**

Reduce Cooling Loads　　　**Visitor Circulation**

TEMPORARY EXHIBITION

Honorable Mention 3

James Law Cybertecture International Holding

Hong Kong

Team

Feisal Noor
Andy Leung
Melvin Pong
Charles Chu

THIS PAGE AND OPPOSITE
Images from competition boards

荣誉奖3

科建国际有限公司（香港）

本页和对页
图片来自于竞赛展示板

SITE PLAN

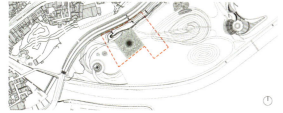

The site is located near the southern tip of Yingge district in New Taipei City, on a reclamation area on the west side of Dahan River. The site is near the Yingge Railway Station and the Yingge Ceramics Museum.

PROGRAMS

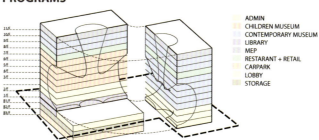

- ADMIN
- CHILDREN MUSEUM
- CONTEMPORARY MUSEUM
- LIBRARY
- MEP
- RESTARANT + RETAIL
- CARPARK
- LOBBY
- STORAGE

ZONING & CIRCULATION

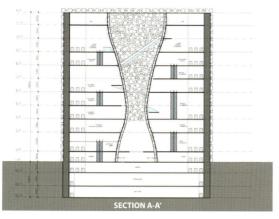

SECTION A-A'

The building is 11 stories high, 77 meters in width, length and height. The program is divided into 3 zones, the contemporary museum, children museum and library, administration zones. From the 2/F lobby visitors can access either one of these zones.

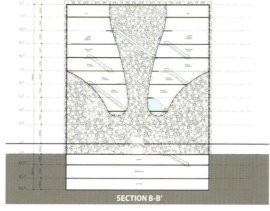

SECTION B-B'

The Northern portion allows access to the 3/F contemporary museum and the special exhibits. The Southern portion allows access to the basement library and administration offices and the 3/F children museum.

James Law Cybertecture® ARUP

2

ART LINK
A Floating Museum Of Nine City Blocks

No.323 Ian Shao+Jr-Gang Chi

This floating museum is a micro-city of great dynamism.

Art Link correlates an art museum with a dynamic city. By synchronizing the major components of the city, the nature, and arts, this 21st-century museum improvises a dialogue between space and culture, while it provides rooms for exhibitions and recreations.

To optimize the connection between the urban and the cultural elements, our proposal significantly redefines and expands the boundary of the site. The new museum encompasses nine "urban blocks," which include the existing Yingge train station and the Yingge agriculture association. In addition, new blocks are offered—courtyard block, passage block, village block, square block, and waterfront block—with the aim of deriving a uniquely enjoyable experience of art and green directly. At the same time, this new urban strategy produces an Axis of Art in the city.

To overcome a 10-meter level change on either side of the Huan-he road that borders the museum site and the city, we propose to expand the museum from above. The result is a sense of buoyancy. As if "floating" above, the museum offers the city a new urban landmark. Consequently mediating between the art and the city, the virtual and the reality, the project at once creates an imaginative panorama of art, green space, and everyday life in the air.

The entrance of the museum will coincide with the entrance passage of the existing Yingge agriculture association. Locally, the perceptive scale will simulate that of the existing street of Yingge. Through the museum circulations, the existing urban fabrics, activities, and the new art programs and spaces will coalesce into a contingent whole. The museum and the city become interchangeable places. One can move effortlessly from the everyday street lives to the art scenes and vice versa, as the Yingge city center becomes the extension of the museum deploying art events in the city. Comparable to the Nolli plan re: public urban space, the museum plan equates to a dynamic urban art plan, as it transforms and grows continually, never coming to an absolute end.

Existing Urban Axis 原有都市軸線

Proposal For New Site 新基地範圍

Linear Urban Interface 綠性城市界面

這是一個漂浮的美術館，一座動態變化的微型城市。
This floating museum is a micro-city of great dynamism.

Roof Garden +57m
Gallery +50m
Library +46m
Admin +42m
Parking +38m

Gift Shop & Book Store

Children Art Museum

Contempor

Figure 1　　　　Existing block : Plaza + Children's Museum

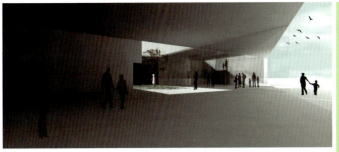

Figure 2　　　　Courtyard block : Galleries

Figure 3　　　　Passage block : Digital media galleries

Figure 4　　　　Village block : Small galleries

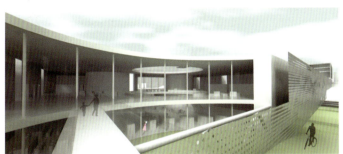

Figure 5　　　　Square block : Large galleries

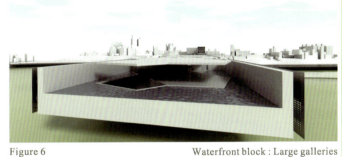

Figure 6　　　　Waterfront block : Large galleries

Figure 7　　　　Museum extending over Huan-he road

Figure 8　　　　View towards river side park

Honorable Mention 4

Ian Yan-Wen Shao
with
JR-Gang Chi,
Ar-Ch Studio
Taiwan

THIS PAGE AND OPPOSITE
Images from competi-
tion boards

荣誉奖4

邵彦文和漆志刚（Ar-Ch建筑工作
室）（中国，台湾）

本页和对页
图片来自于竞赛展示板

9. Roof Garden
10. Library
11. Reading room
12. Lecture Hall
13. Admin.
14. Collection Storehouse

eum

Honorable Mention 5

Ysutaka Oonari
Masamichi Kawakami
101 Design
Japan

THIS PAGE AND OPPOSITE
Images from competition boards

荣誉奖5

欧纳利·雅隆和川之上雅道
（101设计）（日本）

本页和对页
图片来自于竞赛展示板

Cylinders and Rings

The museum is composed by cylinders and rings. The ring is composed of two layers, the user's ring on the exhibition floor and the administrate ring that is under the exhibition floor which also functions as piping space for equipments.

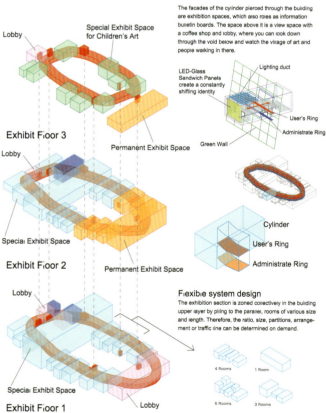

Lobby
Special Exhibit Space for Children's Art

Exhibit Floor 3
Lobby
Special Exhibit Space
Permanent Exhibit Space

Exhibit Floor 2
Lobby
Special Exhibit Space
Lobby
Exhibit Floor 1

The facades of the cylinder pierced through the building are exhibition spaces, which also roles as information bulletin boards. The space above it is a view space with a coffee shop and lobby, where you can look down through the void below and watch the village of art and people walking in there.

LED-Glass Sandwich Panels create a constantly shifting identity

Lighting duct
User's Ring
Administrate Ring
Green Wall

Cylinder
User's Ring
Administrate Ring

Flexible system design
The exhibition section is zoned collectively in the building upper layer by piling to the parallel, rooms of various size and length. Therefore, the ratio, size, partitions, arrangement or traffic line can be determined on demand.

4 Rooms   1 Room
6 Rooms   3 Rooms

The ring composition in which connects each cylinder together has the following advantages.

1: Each exhibition room's ratio, size, and length can be decided according to the cylinder shape.

2: The room with windows for natural light or with no window (open internally open on the outside) can be chosen by arrangement.

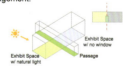

Exhibit Space w/ natural light   Exhibit Space w/ no window   Passage

3: There is a passage between the exhibition room and the exhibition room, and the exhibition room privacy can be kept.

4: The passage between each exhibition rooms will function as a place to remind the former exhibition, and also to switch the feelings preparing for the next following exhibition.

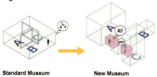

Standard Museum   New Museum

5: The cylinder also acts as a short cut route besides the main route, which will make it able to plan both an exhibit story-line to follow as usual or a new free exhibit.

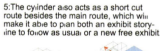

Usual Circulation   Shortcut

6: As each cylinder has their own structure & equipment, it is able to make exhibition routes as using fire or water, or exhibition rooms with art workshops.

×   ○

7: The schedule can be overlapped. By using the 3 meter wide user's circle the administrate circle, it allows the exhibition replacement during the museum open period or showing pieces with a different periods of starting and ending.

Ex A
Ex B
Ex C
Ex D

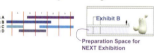

Exhibit B
Preparation Space for NEXT Exhibition

8: As the visiting route of exhibition rooms can be chosen freely, the story of the exhibition can be personal. Whenever you come, there is an opportunity to learn a new art feeling.

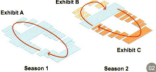

Exhibit B
Exhibit A
Exhibit C
Season 1   Season 2

cylinders / mountain

Art used to be more daily acting as a space in an aristocrat palace life being a part of architecture. Nowadays as the decline of this style, art is becoming more independent from architecture on the same time more non-daily, that almost becoming just an object of a personal collection or even just of an investment. This project focuses on regaining the art to daily life by creating a new relationship between art and architecture, though respecting the art's independence. For a new style of museum representing Taiwan or New Taipei, the museum will be a pile of cylinders that will connect the art inside the exhibition room and the outside city. The cylinders will be covered with plants that will make literally a mountain of art. From inside of the mountain it will be more like a forest that bears fruits of art. It can also be recognized as woods for exhibitions route or a tree for an exhibition room that consist a piece of art individually. The forest will cover a large public space that all together role as an art village.

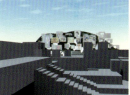

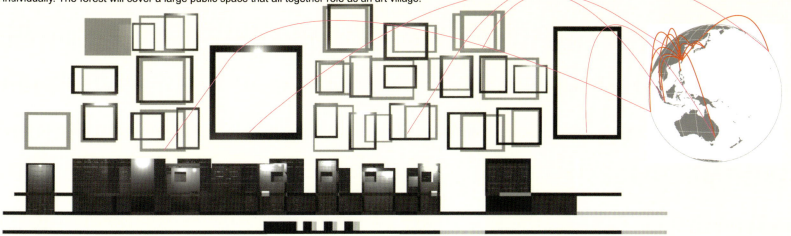

Focusing on the Center

聚焦于中心
法戈的城市填充设计竞赛

Fargo's Urban-infill Design Competition

As the largest city in the U.S. state of North Dakota, Fargo can afford to speculate about a redesign of its downtown core. Considering the state of the U.S. economy, one might question the planning of such an ambitious venture. But, in contrast to the rest of the nation, North Dakota's economy is experiencing boom-like symptoms, supported mainly by the energy and agricultural sectors. Until recently, most outsiders regarded Fargo as a sleepy, northern, small city. Now, with a metropolitan population of 200,000 and growing, the community can think bigger and better. Choosing a design competition for a downtown plan is an interesting move in this direction, even though this was only an ideas competition, and there is no guarantee any of the ideas from this event will be used.

The competition was sponsored by The Kilbourne Group, a progressive local developer specializing in a broad range of urban projects—with concentration on the preservation and reuse of existing structures, as well as new construction. The site of the competition was a city block located in the very center of the downtown, with the focus on creating a multi-use space that could accommodate retail, residential, parking, office and open space for civic use, bringing new energy to the downtown. The idea for a competition arose in 2009 when a dated parking garage occupying one-quarter of the U.S. Bank block was demolished. Rather than see this garage simply replaced by a new parking structure, The Kilbourne Group felt that a competition could generate a more progressive solution for the site.

As an ideas competition for a specific site, open to students as well as professionals, it attracted 160 entries from 23 countries. Because its schedule coincided with the academic semester, a number of studios at schools of architecture—The University of Cincinnati is one example— were able to include it in their curriculum.

Many of the teams chose a similar approach to the site, featuring a large atria as an entrance feature bounded by retail and a highrise.

Of the total prize money of $29,000, there were two first place awards of $10,000 each. One went to "Fargo 365" by a design team from Philadelphia, Pennsylvania, and the other to "Ebb and Flow," designed by Nakjune Seong and Sarah Kuen, two college students from South Lake, Texas. Second place *($5,000) was awarded to the local Helenske Design Group's "Vertical Plain."Many of the teams chose a similar approach to the site, featuring a large atria as an entrance feature bounded by retail and a high-rise. Here one invariably has to think of Berlin's Sony Center by Helmut Jahn. Although this block is considerably smaller than the Sony site in Berlin, lessons can be learned from Sony that are in part implemented here, the most important being multiple penetrations into the site's interior. The competition produced a number of interesting ideas, from parking structures to soaring towers challenging the eye.

A number of very doable ideas resulted, and one can only wonder if the powers that be might engage one of the winners to provide Fargo with a new downtown look. –SC

作为美国北达科他州最大的城市，法戈打算对市中心进行重新设计。考虑到美国经济现状，人们也许会质疑这一计划是个大冒险。但是，与美国其他地区相比，北达科他州正依靠能源和农业蓬勃发展。如今，外地人总是认为法戈是一个沉睡的北部小城。现在，法戈已经拥有200,000人口，正在不断地变大、变好。选择设计竞赛来进行市中心规划是朝向前进的重要一步。尽管这只是一个概念竞赛，或许以后这些方案将被实现。

竞赛由吉尔伯恩集团赞助。该集团是当地一家先进的开发商，致力于各种规模的城市项目，既有现有建筑的保护和再利用，又有新项目的建造。竞赛场地是位于市中心的一个街区，聚焦于打造多功能空间，包含零售、住宅、停车场、办公室和公共开放空间等设施，为市中心注入新活力。竞赛的概念源于2009年，正当一个占据美国银行街区四分之一面积的过时车库被拆除之际。相对于让车库被新的停车设施所取代，吉尔伯恩集团认为一次竞赛将为该场地带来更加先进的解决方案。

作为一次在指定场地上的概念竞赛，竞赛面向学生和专业人士开放，供吸引了来自23个国家的160件参赛作品。由于竞赛日程与学校学期相符，许多建筑学院的研究室（例如辛辛那提大学）都将它纳入课程安排之中。

竞赛的总奖金为29,000美元，两个一等奖得主分获10,000美元。一个被授予了来自费城的设计团队所设计的"法戈365"，另一个则由南湖城的两名学生所设计的"潮起潮落"获得。二等奖（5,000美元）由当地海伦斯基设计集团的"垂直平原"获得。

许多团队都选择了类似的设计方式，以一个巨大的前庭作为入口特征，四周环绕着零售空间和高层建筑，让人们不得不想起由赫尔姆特·扬设计的柏林索尼中心。尽管这一街区要比索尼中心的场地小很多，索尼中心的经验教训仍然值得借鉴。竞赛打造了一系列的有趣概念，从停车设施到高楼大厦，一应俱全。竞赛产生了许多可行的概念，优胜者的设计将会为法戈带来一个全新的市中心景象。——斯坦利·科利尔

First Place (2)

"Ebb and Flow"
Nakjune Seong
Sarah Kuen
South Lake, Texas

LEFT
Birdseye view and illustrations from
the competitions board

FARGO, ND
LATITUDE 46° 52' 37.66" N
LONGITUDE 96° 47' 23.50" W
ELEVATION 905 FEET

ELEVATION 43.5°
AZIMUTH 285.4°

NATURAL VENTILATION IN ATRIUM DURING SPRING & FALL

ELEVATION 66.6°
AZIMUTH 323.3°

NATURAL VENTILATION IN ATRIUM DURING SUMMER

ELEVATION 19.7°
AZIMUTH 247.5°

COLLECTING RADIANT HEAT IN ATRIUM DURING WINTER

SPRING AND FALL

OPEN ROTATED AND LOCKED WALL DOORS
RETRACTING GLASS WALL TO ENLARGE OPENING
INDOOR BECOMES OUTDOOR.

WINTER

CREATE MOST SUSTAINABLE ENVIRONMENT IN AN ENVELOPE

OUTDOOR SPACE BECOME INDOOR SPACE.
THE WALL DOORS ARE CLOSED
AND CREATE GREEN HOUSE EFFECT.

一等奖（2）

"潮起潮落"
那俊・成和萨拉・权（得克萨斯州，南湖城）

左图
鸟瞰图和竞赛展板插图

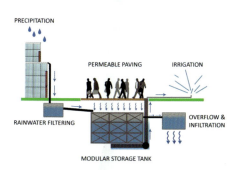

PRECIPITATION

PERMEABLE PAVING

IRRIGATION

RAINWATER FILTERING

OVERFLOW &
INFILTRATION

MODULAR STORAGE TANK

RAINWATER HARVESTING AND REUSE CONCEPT

MAIN ENTRANCE PERSPECTIVE

ebb

SITE CONTEXT

RENAISSANCE ZONE

VISION STATEMENT: The Red River Town Plaza reflects the community's desire to establish a *DESTINATION* for locals, young professionals, college students, and visitors alike. Boasting a unique blend of retail, residential, and office spaces, Red River Town Plaza provides a wonderful live, work, and play lifestyle. Similar to the Red River nearby, the plaza offers paths and spaces of *HIGH ENERGY* and entertainment and an upbeat, urban residential community. In addition, meandering paths along the water feature and gathering places within the plaza offer opportunities to *SLOW DOWN AND RELAX*, imitating the Red River's slow steady channels and wide, calming bends.

A unique feature of the Red River Town Plaza, the glass atrium remains open during the warmer seasons, encouraging visitors to *FLOW* in and out of the spaces, *CIRCULATE* through the Broadway area, or settle in at one of the community events being hosted in the plaza. During the winter, the retracting glass panels of the atrium remain closed, allowing residents and visitors to experience the warm and charming downtown Fargo experience, even when the weather outside may be less hospitable.

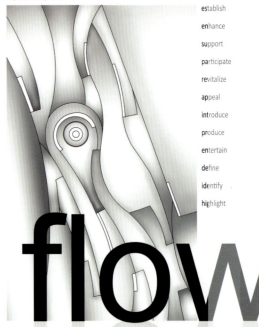

establish
enhance
support
participate
revitalize
appeal
introduce
produce
entertain
define
identify
highlight

flow

CURRENT LAND USE PLAN

Commercial
Office
Government
Theater
Hotel
Residential
Parking
Parks/Open Space

HISTORICAL DISTRICT + HISTORIC BUILDINGS

SUN ANGLE AND WIND DIRECTION

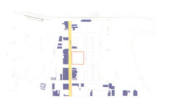

BROADWAY COMMERCIAL CORRIDOR

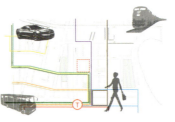

MULTI-MODAL TRANSPORTATION

SKY WALK CONNECTION AND VIEWS INTO THE SITE

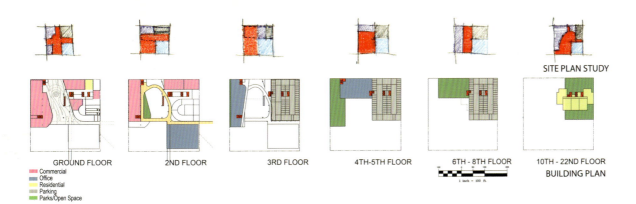

SITE PLAN STUDY

| | | | | | |
| GROUND FLOOR | 2ND FLOOR | 3RD FLOOR | 4TH-5TH FLOOR | 6TH - 8TH FLOOR | 10TH - 22ND FLOOR |

BUILDING PLAN

1 inch = 100 ft.

■ Commercial
■ Office
■ Residential
■ Parking
■ Parks/Open Space

3RD AVENUE NORTH

STREET VENDOR
RETAIL
CONDO LOUNGE
MEDIA TOWER & FOUNTAIN
RETAIL
RETAIL
MECH.
IN-DOOR GARDEN ART GALLERY
CORRIDOR ABOVE
OUTDOOR DINING PATIO
UP
BROADWAY
5TH STREET NORTH
2ND AVENUE NORTH

PLANNED AREA
RETAIL : 35,000 SQ FT
HOUSING : 120,000 SQ FT
OFFICE : 48,000 SQ FT
VEHICLES : 450 CARS

ENTRANCE HALLWAY CORRIDOR

MEDIA TOWER RETAIL ART GALLERY

DINING PATIO FOUNTAIN SEATING

OUTDOOR LANDSCAPE CONTINUES INTO INDOOR SPACE AND BECOMES INTERIOR. THE BODER BETWEEN INDOOR AND OUTDOOR IS DESOLVED.

ENTRY PLAZA 1
WATER FEATURE & SEATING 2
POA PRAIRIE 3
FARGO KNOLL 4
STAIRWAY ACCESS 5
RETRACTABLE ATRIUM DOORS 6
DINING PATIO 7
CABLE & VINE COLUMNS 8
ESCALATOR 9
SEATINGS 10
REVOLVING DOORS 11
PEACE GARDEN ENTRY PLAZA 12
RESIDENTIAL TOWER ENTRY 13
HOTEL ACCESS 14
EXISTING US BANK BUILDING 15

RED RIVER TOWN PLAZA PLAN

1 inch = 20 ft.

First Place (2)

"Ebb and Flow"
Nakjune Seong
Sarah Kuen
South Lake, Texas

The main attraction of the **"Ebb and Flow"** entry was a generous glass atrium. Its retractable entrance could remain open in warmer weather and closed in the winter. A winding path through a plaza to the entrance from the main street was a reference to the nearby Red River. Not only was the site highly visible from the southeast corner, where the main intersection is located; pedestrians can easily enter the site from three different points on all the surrounding streets. -SC

RESIDENTIAL
OFFICE GARAGE
RETAIL

DOWNTOWN FARGO BUILDING SECTION,

16' 20' 80' 12'

BROADWAY AVE. STREET SECTION: LOOKING SOUTH
SCALE 1"=20'-0"

VERTICAL CIRCULATIONS
3RD FLOOR CORRIDOR
2ND FLOOR CORRIDOR
ESCALATOR
SKY WALK

HORIZONTAL AND VERTICAL CIRCULATION

ENTRY PLAZA PERSPECTIVE

一等奖（2）

"潮起潮落"
那俊·成和萨拉·权（得克萨斯州，南湖城）

"潮起潮落"的主要吸引力来自于宽敞的玻璃中庭。它的可伸缩入口能够在冬季闭合，温暖时开放。曲折的小路通过广场通往街面的主入口，与紧邻的红河有异曲同工之妙。场地的东南角十分显眼，那里设置着十字路口；行人可以通过周边三个方向的街道进入场地。——斯坦利·科利尔

FARGO365

A CIVIC SPACE

First Place (2)

"Fargo 365"
David Witham,
Doug Meehan,
Anna Ishii and Hannah
Mattheus–Kairy
Philadelphia, PA

LEFT AND OPPOSITE PAGE
Illustrations from competition
boards indicating seasonal
changes

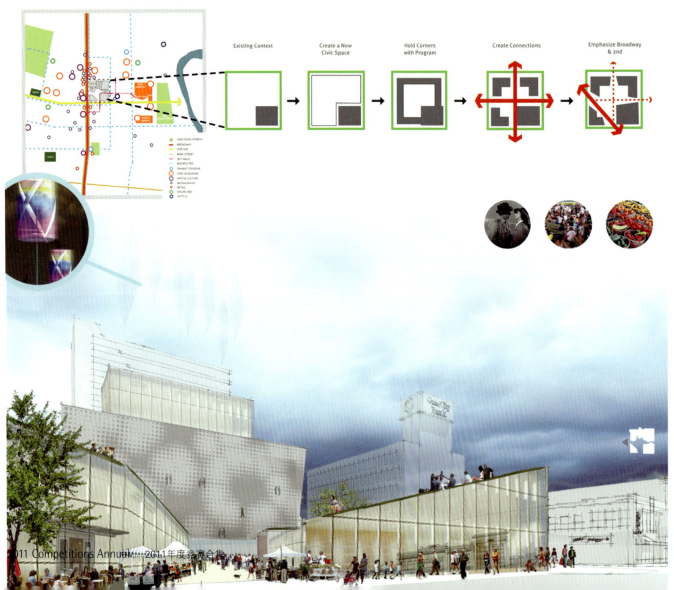

| Existing Context | Create a New Civic Space | Hold Corners with Program | Create Connections | Emphasize Broadway & 2nd |

一等奖（2）

"法戈365"
大卫·维特汉姆；道格·米汉；安娜·石井和汉娜·马修斯–凯利（宾夕法尼亚州，费城）

左图和对页
竞赛展板插图展示了季节变化

"法戈365"凸显了场地在一年之中的多种用途。设计师通过将一个低调的三角形元素设在角落，营造出广场的半封闭造型。广场上方的绿色屋顶逐渐下降至室内。分别位于两条相邻街道上的两个主入口凸显了广场的景色，令广场享有了一个强烈的焦点，吸引着人们前来。此外，半封闭式广场保证了内部私密感，同时又不为外部行人设置障碍。除了高层结构之外，周边的建筑都采用了绿色屋顶，既具保护性又环保节能，特别适合北达科他州寒冷的冬季。

Fargo 365" emphasized the multi-purpose use of the site throughout the year. Here the plaza's semi-interior form was created by locating a modest triangular element at the corner, featuring a green roof, gradually sloping down to the interior. Punctuated by two main entrances from the two adjoining streets, the partially hidden plaza received a strong focus, drawing people in by virtue of the functional nature of the site. Moreover, by partially enclosing the plaza, a certain level of intimacy is present, without creating a barrier to the outside pedestrian. With the exception of the high-rise structure, the buildings all were adorned by green roofs, certain added protection—and an energy saver—in view of North Dakota's severe winters. -SC

FOR EVERY SEASON

IN CELEBRATION OF CLIMATE + CULTURE + ART

WINTER

SUMMER

INTERACTIVE LIQUID CRYSTAL SCREEN

Vertical Plain

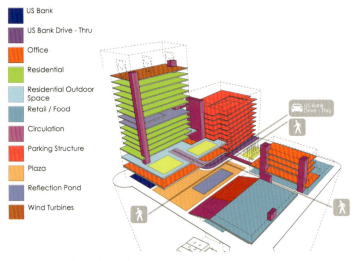

US Bank

US Bank Drive - Thru

Office

Residential

Residential Outdoor Space

Retail / Food

Circulation

Parking Structure

Plaza

Reflection Pond

Wind Turbines

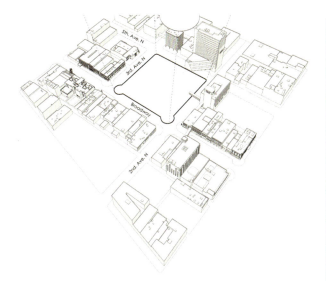

二等奖
"垂直平原"
海伦斯基设计集团（北达科他州，法戈）

"垂直平原"主要特点在于高层结构的处理象征了"平原"元素。通过融入太阳能电池板和风力发电机，该设计具有高度的节能特征。然而，要不是设计具有北达科他州平原的象征意义，这个设计可能早就已经被淘汰了。——斯坦利·科利尔

Vertical Plain

The metaphor driving the design was seeking something inherently of North Dakota, of the region, a context of the place, and not necessarily limited by historical precedent or the building immediately neighboring the site. The context arrived at is the Plains. The plains, in their simple, yet powerful horizontal nature, are what sustain, anchor, nurture and shelter us. The goal was to take a cross section of the region and arrange it vertically, expressing symbolically what we hold dear. The most obvious element of the design solution is the literal plane, which articulates the various programmatic requirements of the tower, folds over to create the plaza, and folds again into the sloped greenspace and sheltering canopy. A more subtle parallel is found on the buildings' skins, which consist of horizontal photovoltaic (solar powered) glass bands. The bands are arranged in such a way as to allow light and air into the building, while visually appearing as tilled soil, some stripped away, some remaining. The final reference to the plains looks to the present, and more importantly the future, of further utilization of the plains; literally in the form of the wind turbines on the uppermost portion of the building. This functional element, and symbolic reference, hopes to symbolize an optimistic future and increased relevance for the region in the future.

Second Place
"Vertical Plain"
Helenske Design Group
Fargo, North Dakota

"Vertical Plain," was mainly about the treatment of a high-rise structure which was supposed to symbolize elements of the 'plain.' By including solar panels and wind turbines in the mix, this design strived to give the impression of a highly energy-efficient structure. However, without the narrative on the symbolic connection to the North Dakota plain, this would probably be lost on the casual observer. -SC

A. MEDIUM DENSITY RESIDENTIAL BLOCK

A medium density residential apartment building with retail at grade level, with a podium for the top four floors. The building is low so it does not cast a shadow on the public space to the north.

Floor	Use	Sq Ft
-1	Parking	13,000 (*47)
Ground	Retail	16,320
2	Residential	14,260
3	Residential	10,070
4-7	Residential	7,400

- ■ Residential
- ■ Retail / Cafes + Bars
- ■ Office Commercial
- ■ Parkade

B. HIGH DENSITY RESIDENTIAL TOWER

A high density residential tower, designed to be the second tallest building in the city with smaller apartment sizes, located in the the north side of the public space. Retail frontages wrap the building at grade level.

Floor	Use	Sq Ft
-1	Parking	30,000 (*109)
Ground	Retail	6,950
2-14	Residential	5,720

C. INTEGRATED PARKADE-OFFICE BLOCK

A parking structure with entrance and exit off of 5th Street with an integrated office building above. Retail uses are located on the ground floor, with unsightly parkade facades above. A green roof sits on top for use by office workers.

Floor	Use	Sq Ft
-1	Parking	21,660 (*78)
Ground	Retail / Parking	10,990/3,000 (*11)
2	Retail / Parking	10,990/3,000 (*11)
3-5	Parking	14,390 (*156)
6-7	Commercial Office	10,990
8-12	Commercial Office	8,570

*275 sq ft per parking stall incl. turning and aisles
This proposal actively suggests 20% less parking spots than suggested, which should be complemented with a public transit analysis to promote ridership.

PROPOSED BUILT FORM + SPECIFICATIONS

STREETSCAPE ALONG BROADWAY

It is important to ensure that buildings of heritage value are protected within Fargo, however, new buildings should not attempt to recreate older architectural styles; instead, they should seek to complement existing buildings in scale, massing, and proportion. As such, buildings facing Broadway have a 2-storey podium to create a sense of cohesion along the historic street.

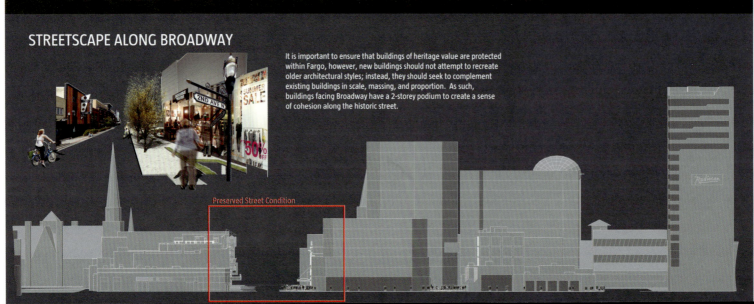

Preserved Street Condition

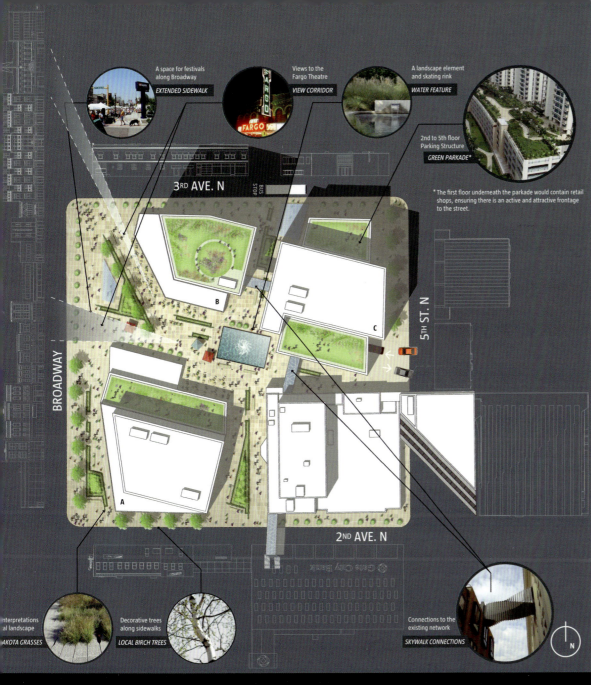

A space for festivals along Broadway
EXTENDED SIDEWALK

Views to the Fargo Theatre
VIEW CORRIDOR

A landscape element and skating rink
WATER FEATURE

2nd to 5th floor Parking Structure
GREEN PARKADE*

* The first floor underneath the parkade would contain retail shops, ensuring there is an active and attractive frontage to the street.

3RD AVE. N
BUS STOP

5TH ST. N

BROADWAY

B

C

A

2ND AVE. N

...interpretations
...al landscape
...AKOTA GRASSES

Decorative trees along sidewalks
LOCAL BIRCH TREES

Connections to the existing network
SKYWALK CONNECTIONS

N

UBLIC REALM + PEDESTRIAN COMFORT

EALING WITH SHADOWS

...ril 1
...am

5 pm

Spring and Summer shadows do not reach the majority of the public space during all times of the day, and Broadway is only shaded in the morning.

...ober 1
...am

5 pm

While Fall and Winter shadows are stronger, most of the public space still remains unshaded and comfortable, while Broaday is also unshaded during the morning.

A WINTER MICROCLIMATE

The design and shape of the buildings surrounding the public space were made to shield the squarer from winter winds, creating a space that is pleasant for people to enjoy, even during the tough winter months when people are generally indoors. This microclimate simulation shows how the public space is shielded from harsh winds.

Winter Winds in Fargo

Most winter winds are from a Northwesterly or Southeasterly direction, with the strongest ones coming from the North. The taller buildings in the north will shield the public space, while still allowing for views of Broadway.

COLDER — WARMER

PROGRAM

NEW BUS STOP

FESTIVAL SPACE OFF BROADWAY

ACTIVE FRONTAGE

ACTIVE FRONTAGE

CENTRAL PLAZA

RETAIL STREET

LANDSCAPED PEDESTRIAN CONNECTOR

ACTIVE FRONTAGE

Central Plaza Cafes and Bars in the warmest, unshaded parts of the public space surrounding a water feature.

Festival Space A space for the Farmer's Market, Community Art Projects, Film Festival, etc.

Retail Street Retail shops along Broadway in a modern interpretation of 'Main Street'

New Bus Stop A new Bus Stop is reccomended along 3rd Ave for new residentis and visitors.

Connector A landscaped pedestrian connection links the US Bank building with the plaza and other downtown destinations.

荣誉奖

"更加百老汇"
罗素·柯林（加拿大，卡尔加里）

本页和对页
竞赛展示板

Honorable Mention

"More Broadway"
Russel Collin
Calgary, Alberta

THIS PAGE AND OPPOSITE
Competition boards

AMTRAK
RED RIVER
X
MINNESOTA
NORTH DAKOTA
BNSF Railroad
MAIN AVE

DOWNTOWN AREAS
Renaissance Zone
City Parks
Mature Neighborhoods

DESTINATIONS
1 Amtrak Station
2 City Hall / Civic Centre
3 Fargo Theatre
4 Transit / Bus Depot
5 Medical Center
6 Plains Art Museum
7 Visitor Center
8 Public Library

TRANSPORT
P Parking
T Transit Hub
Bus Route
Train Route

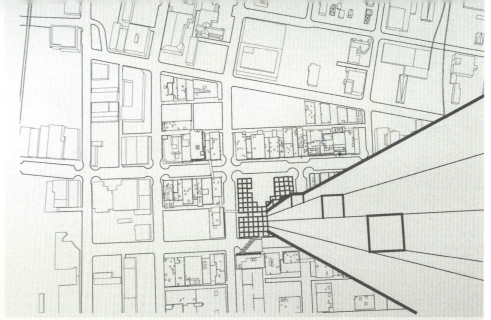

荣誉奖

"地标"

泰·格里夫（新墨西哥州，圣达菲）
竞赛展示板的插画和图纸

Honorable Mention

"Icon"
Ty Greff
Santa Fe, New Nexico

THIS PAGE AND OPPOSITE
Competition board with
illustration and plan

Icon or: How Fargo became the center of the architectural universe

Honorable Mention

荣誉奖

"Seed"
Ted L. Wright
Phoenix, Arizona

"种子"

泰德・L・莱特（亚利桑那州，凤凰城）

本页和对页
竞赛展示板

THIS PAGE AND OPPOSITE
Competition boards

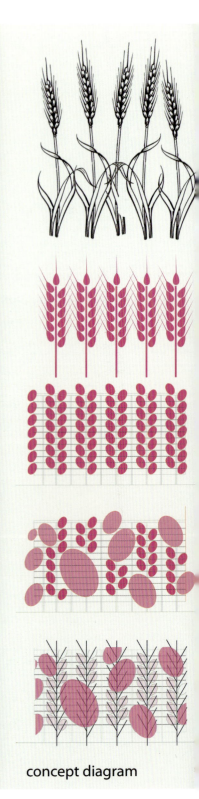

concept diagram

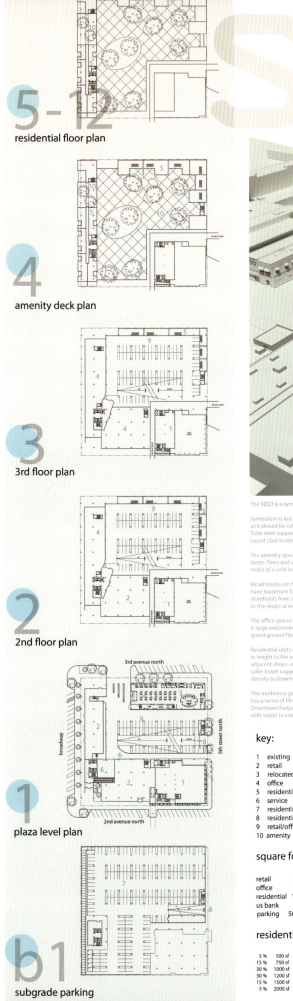

5-12
residential floor plan

4
amenity deck plan

3
3rd floor plan

2
2nd floor plan

1
plaza level plan

3rd avenue north

broadway

5th street north

2nd avenue north

b1
subgrade parking

The SEED is a symbol of promise and potential. The eternal cycle of renewal and bounty begins with one seed. No other industry or way of life relies on these truths more than Agriculture.

Symbolism is key to understanding cultures and geography. North Dakota, in particular the Red River Valley, has some of the richest soil in the world. Agriculture is a large part of this regions identity and should be celebrated. Agriculturally related symbols are used throughout this design. The seed or kernel with its familiar simple geometry shapes the residential balconies and amenity spaces. Tube steel supports are integrated into the buildings cladding similar to the shaft of a wheat stock. Smaller tubes spring at angles from the stalk mimicking beards of barley or wheat. The white translucent clad residential tower glows at night as if to welcome the community similar to the way a light pours out of the window suggesting welcoming warmth on a cold winter's night.

The amenity spaces are plentiful throughout the design which borrows the symbolism mentioned above. On the façade, elliptical seed shaped openings frame gathering areas common to all the residents. Trees and vegetation grow on elevated terraces above the street for all the residents to enjoy. Two of the three are enclosed ensuring a lush warm environment so welcomed by those in the midst of a cold long winter.

Retail resides on floors one and two for easy off the street access ensuring tenants will have maximum foot traffic. The residential tower overhangs the retail which shelters the storefronts from inclement weather; but also provides a sense of smaller scale which relates to the shops across the street.

The office spaces are situated two floors above retail and below the residential tower. A large welcoming lobby is accessed by the multi level parking garage or from the grand ground floor lobby.

Residential units occupy the upper most floors forming a twelve story tower which compares in height to the adjacent Radisson Hotel. The shorter podium program elements relate adjacent shops, intentionally designed not to overwhelm the smaller scale context yet the taller tower suggests an element of vision and prosper introducing the right amount of density to downtown.

The multistory garage is skinned by live, work or play spaces ensuring the streetscape always has a sense of life and never a blank garage.
Downtown Fargo is ripe with un-harvested bounty and promise. But it first takes individuals with vision to energize and plant the SEED.

key:

1 existing
2 retail
3 relocated us bank
4 office
5 residential
6 service
7 residential parking
8 residential ramp
9 retail/office parking
10 amenity

square footage

retail	38,000 sf
office	32,000 sf
residential	100,000sf
us bank	10,000 sf
parking	502 spaces

residential unit sizes

5 %	500 sf
15 %	750 sf
30 %	1000 sf
30 %	1200 sf
15 %	1500 sf
5 %	2000 sf

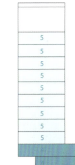

stacking diagram

BROADWAY MARKET COMMO

**People's Choice
1st Place Award**

**"Broadway
Market
Commons"
Cornell Design
and Planning
Group**
Kevin Gill
Dasha Mikic
Chuijing Kong
Heather Blaikie
Robert Krumhansl
Ithaca, NY

THIS PAGE AND OPPOSITE
Competition boards

最受民众欢迎奖一等奖

"百老汇市场"
可奈尔设计规划集团（纽约
州，伊萨卡）

本页和对页
竞赛展示板

DESIGN FEATURES
- INDOOR/OUTDOOR MARKETPLACE
- PUBLIC PLAZA
- PEDESTRIAN ORIENTED MIXED USE DESIGN
- ON-SITE STORMWATER MANAGEMENT SYSTEM

AERIAL PERSPECTIVE

STORMWATER SYSTE

Precipitation Storage
Controlled Rainwater Release
Evaporation
Large Gravel/ Cobble

Precipitation Storage
Controlled Rainwater Release
Water Reuse
Building Insulation
Glare Reduction
Vegetation

Overflow to
Storm Sewer

BioRetention
Surface Water Storage
Water Quality Filtration
Controlled Stormwater Relea
Groundwater Recharge

LIVE

84,000 SF NEW HOMES
+ 13,500 SF CONVERTED LOFTS

97,500 SF RESIDENTIAL

WORK

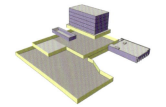

44,800 SF NEW CLASS A OFFICE
+ 7,800 SF RETAINED OFFICE
52,600 SF OFFICE
≈ 375 STRUCTURED PARKING SPACES

PLAY

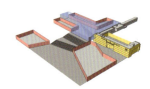

23,000 SF RETAIL
21,600 SF OPEN SPACE
12,000 SF INDOOR MARKETPLACE
6,500 SF COMMUNITY CENTER

MIXED USE

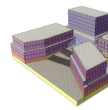

RETAIL
RESIDENTIAL
OFFICE
PARKING
COMMUNITY CEN

SITE PLAN

3RD AVE N.

MARKET COMMONS

INDOOR MARKETPLACE

N. BROADWAY DR

5TH ST N.

2ND AVE N.

PERSPECTIVES

MARKET COMMONS

PEDESTRIAN STREETSCAPE BORDERING MARKET COMMONS

INDOOR MARKETPLACE AT NIGHT

Ambitious Plans for a New Belgrade
The Centre for the Promotion of Science

by Ted Sandstra

新贝尔格莱德的宏伟蓝图
科学推广中心

泰德·桑德斯特拉

Background

在重建过程中，塞尔维亚首都贝尔格莱德——正尽量远离它过去对民族主义的狂热。日前，塞尔维亚正在申请进入欧盟，并以促进文化和经济进程为首要任务。作为新指导方针的一部分，市政府新近举办了两次城市项目竞赛，旨在将贝尔格莱德塑造成一个不断扩张而又注重公民和国家形象的城市。因此，塞尔维亚政府举办了科学推广中心竞赛、2011年哈拉·贝通滨水中心竞赛和卡莱梅格丹公园竞赛。

之所以将科学推广中心作为首次国际竞赛的原因在竞赛提纲中得到了明确的解释：

"科学中心激发好奇心并且帮助儿童了解科学。在以知识为基础的社会中，一个现代科学中心在科学文化宣传和强化研究工作上起到了中心作用。这不仅事关年轻一代，也关乎成年人。"

In the process of reinventing itself, the Serbian capital of Belgrade, once the political center of Yugoslavia before its dissolution into a number of smaller nation states, is steering away from its obsession with its nationalist past, and, with its application for entry into the European Community, is promoting cultural as well as economic progress as its top priorities. As part of this new direction, the City staged recent competitions for two city projects, bringing attention to the needs of Belgrade as a large city that not only continues to grow, but is soliciting ideas that speak to a civic and national identity that is intended to redefine significant historical locales within its geography. Thus, in the tradition of the Grande Projets, the Serbian administration announced competitions for a Center for the Promotion of Science; and also the Hala Beton Waterfront Centre 2011 and Kalemegdan Park on the Danube, the former located in Block 39 of the New Belgrade plan, the latter located just across the Sava River from it.

New Belgrade was originally designed in the 1950s as the connective tissue between two cities - Belgrade to the east and Zemun to the west. Belgrade was the furthest point west of the Ottoman Empire while Zemun represented the Austrian-Hungarian empire to the west. By finding a way to bind these two cities with a new, modern housing district, Tito s regime intended to indicate a direction forward for a nascent nation.

Why the city regarded a new Centre for the Promotion of Science, as one of its top priorities for an international competition, is clear from the competition brief:

'Science Centers inspire curiosity and support learning about science from early ages. In the area of knowledge-based societies a modern science centre can play a central role in dissemination of scientific culture and the strengthening of research, not only for young generations, but also for adults.'

First Prize

Wolfgang Tschapeller
ZT GmbH Architekten
Vienna, Austria

By locating the program above the street with remarkable structural simplicity and clarity, one can surmise that this approach caught the attention of the jury over other, less successful attempts to reach for height and visibility in an urban landscape dedicated to the automobile. Elevating the program also frees up space for the development of additional amenities for pedestrians at street level.

Raising the exhibition area above the ground plane, according to the author, follows the principles of the Athens Charter as set out by CIAM (International Congress of Modern Architecture). This ties the design to the history of the site, which was conceived at a time when the Athens Charter held sway over urban planning.

The criteria for the jury s evaluation of the winner are clear from its concluding statement:
Yet even as the building s form appears radical, the construction is simple, straightforward, well-considered and well-calculated. The Jury has come to the conclusion that this project precisely fits both the requirements, and aspirations, for the proposed institution, as well as for the city in which it will provide a new, welcome landmark. -TS

该设计脱离地面，具有非凡的结构纯粹感和明晰感，拥有比其他设计更成功的高度和视觉效果。让建筑脱离地面同时还释放了地面用来开发其他人行设施的空间。

根据设计师所说，让展览区域悬浮于地面之上遵循了由现代建筑国际大会所制定的《雅典宪章》。这让设计与场地的历史紧密相连，在场地设计时《雅典宪章》在城市规划中起到了统治地位。

以下是评审委员会对该项目的评语：

尽管建筑的外形十分激进，它的结构还是十分简单、直接，经过了精心规划和计算。评审委员会一致认为该项目完全符合竞赛要求，将为城市提供一个全新的地标。——泰德·桑德斯特拉

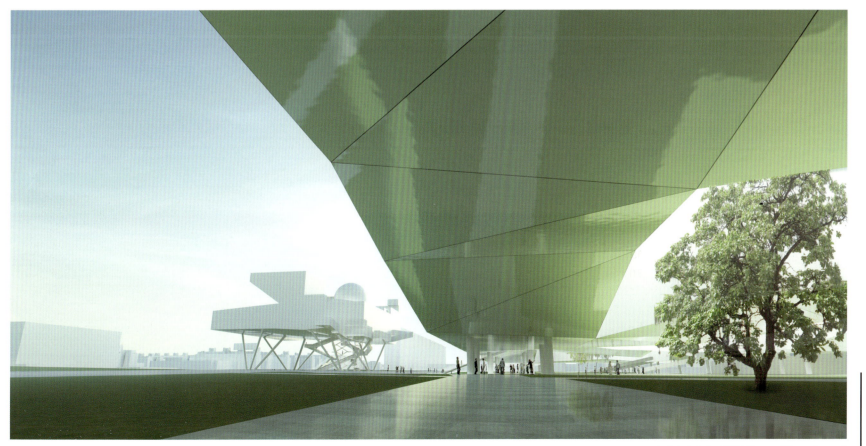

一等奖

沃尔夫冈·沙佩勒尔建筑事务所（奥地利，维也纳）

右图
结构层次
下图
从下方看建筑设计
对页，上图
中心细部
对页，下图
细部特写

First Prize

Wolfgang Tschapeller ZT GmbH Architekten
Vienna, Austria

RIGHT
Layers
BELOW
View underneath
OPPOSITE PAGE, ABOVE
Centre detail
OPPOSITE, BELOW
Close-up detail

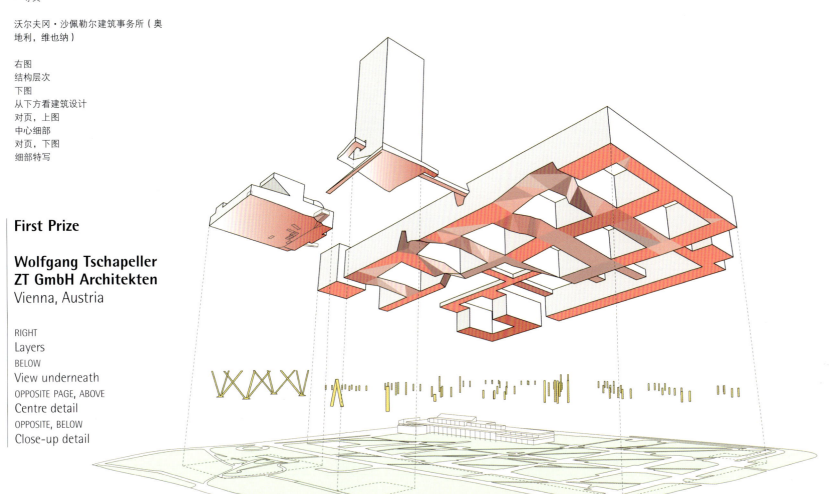

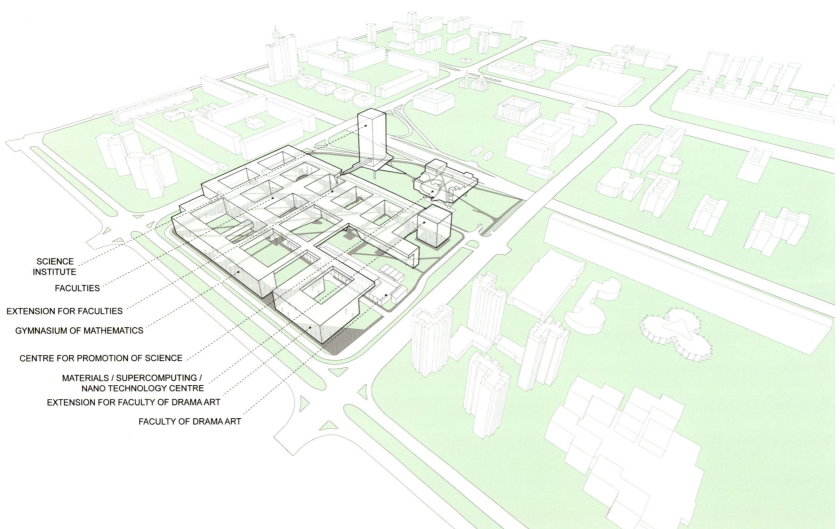

SCIENCE
INSTITUTE

FACULTIES

EXTENSION FOR FACULTIES

GYMNASIUM OF MATHEMATICS

CENTRE FOR PROMOTION OF SCIENCE

MATERIALS / SUPERCOMPUTING /
NANO TECHNOLOGY CENTRE

EXTENSION FOR FACULTY OF DRAMA ART

FACULTY OF DRAMA ART

一等奖

沃尔夫冈·沙佩勒尔建筑事务所（奥地利，维也纳）

右图和下图
剖面图
对页，上图
中心入口
对页，下图
轴测图

First Prize

Wolfgang Tschapeller
ZT GmbH Architekten
Vienna, Austria

RIGHT AND BELOW
Sections
OPPOSITE PAGE, ABOVE
View to arrival at Centre
OPPOSITE, BELOW
Axonometric

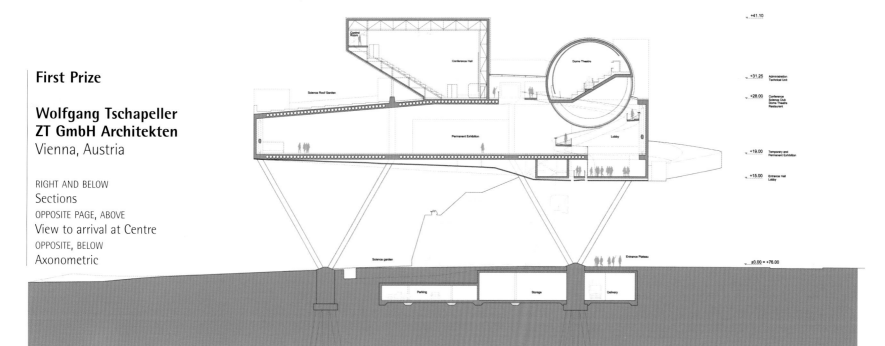

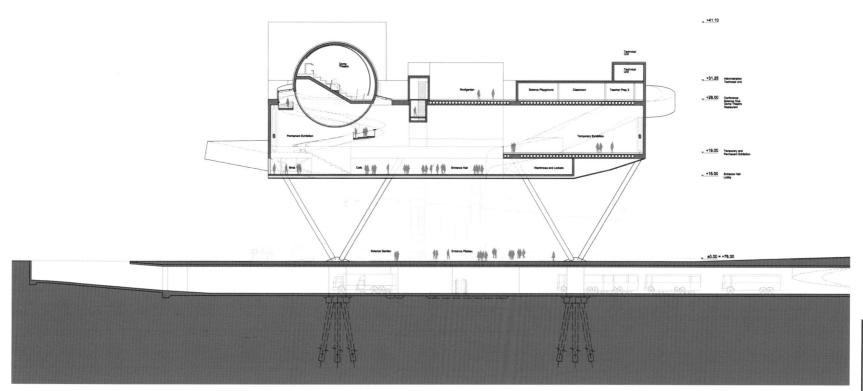

First Prize

Wolfgang Tschapeller
ZT GmbH Architekten
Vienna, Austria

THIS PAGE
Elevations (from top to bottom):
Northeast
Northwest
Southeast

OPPOSITE PAGE, ABOVE
Structural system
OPPOSITE, BELOW
Grade level plan

一等奖

沃尔夫冈·沙佩勒尔建筑事务所（奥地利，维也纳）

本页
立面图（由上至下）
东北立面
西北立面
东南立面

对页，上图
结构系统
对页，下图
地面层规划

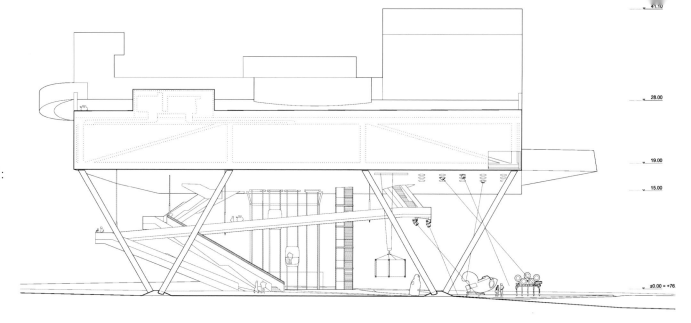

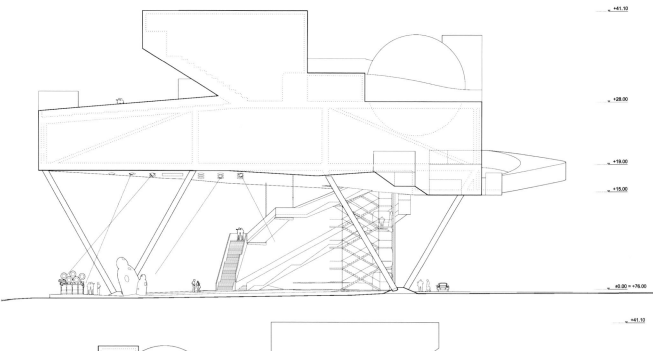

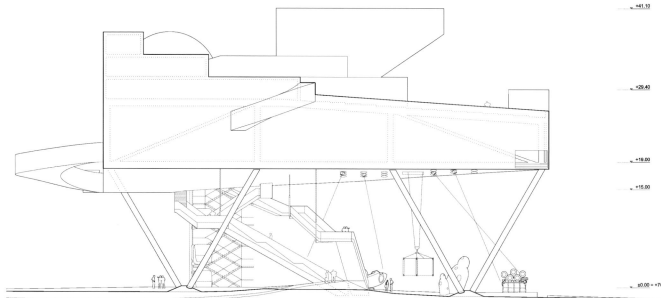

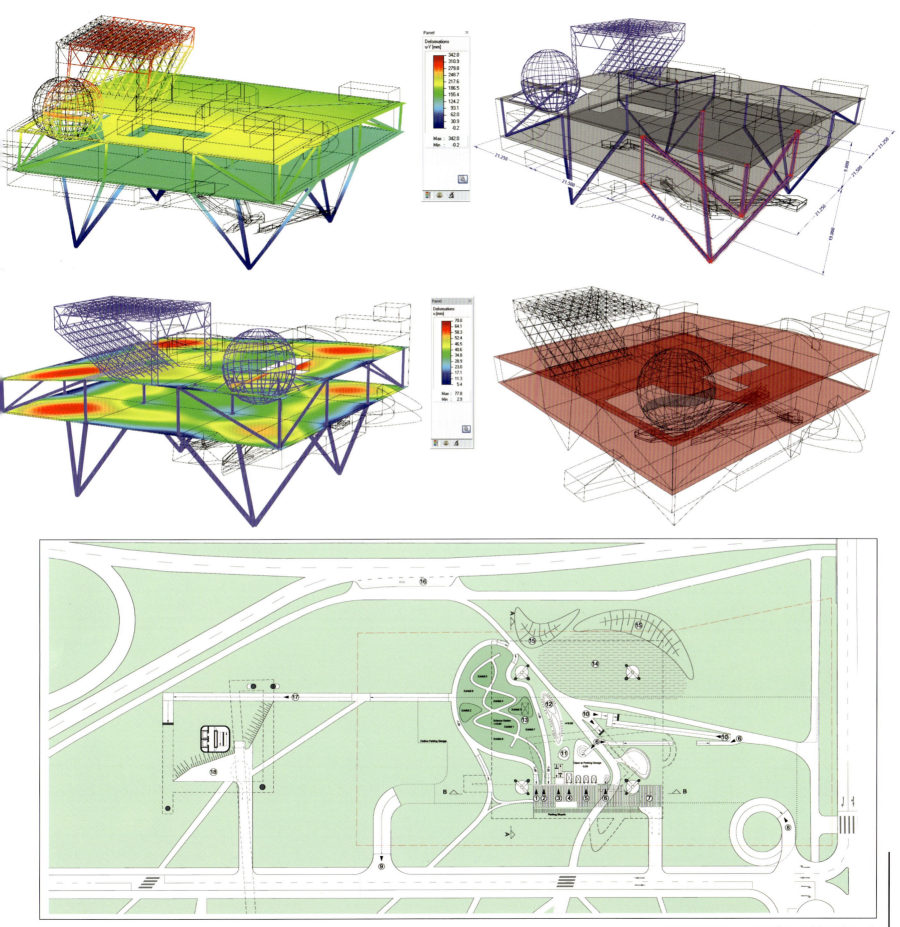

ENVIRONMENTAL CONCEPT

PHOTOVOLTAIC CELLS
To generate electricity to supplement the power for artificial lighting.

OPERABLE SUN ROOMS
The Sun Room is able to be opened in order to enhance ventilation with natural ventilation by solar chimney effect in mid-summer season. Room in winter acts as greenhouse, solar warmed-up using the Sun Room. It will be able to transform and to conserve its space in its night-time lighting to glow the warmer shapes.

RAIN WATER HARVESTING
To reduce the water consumption by collection of the rain water using the rain water for grey water use. (hydrographic scale).

RADIANT HEATING
Radiant heating pipes are on the floor model to increase the comfort range by heating the integrated level with bring up the shrinking over, energy part with the change saving system, reflect the efficiency without damaging heating system.

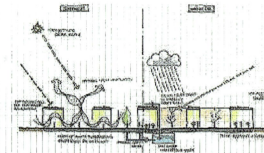

DAYLIGHT FACTOR COMPARISON TO OPTIMIZE DAYLIGHT
In terms of optimizing the lighting level of the premises, the natural contribution of daylight factor to provide the interior spaces of the various resources with daylight coefficients to reduce the daylighting consumption by artificial lighting during the day time. The comparative results results between daylight factor with and without sun room, that the daylight factor with and without sun room with rooms of the various resources.

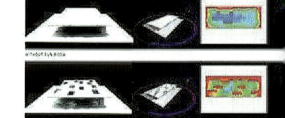

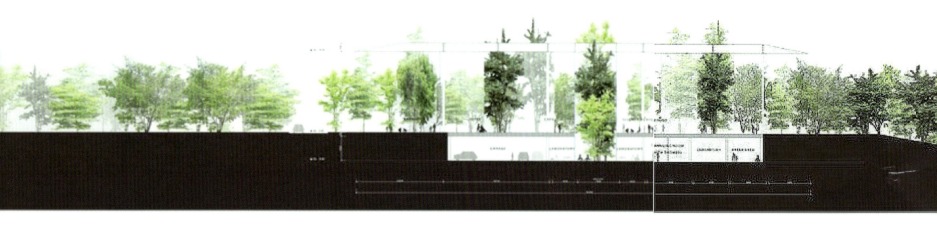

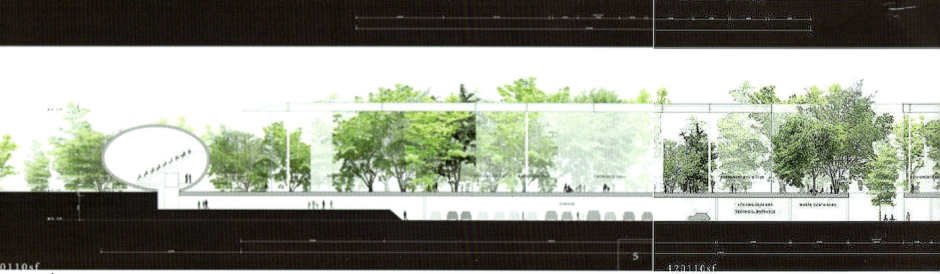

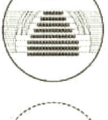
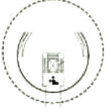
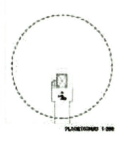

URAL CONCEPT

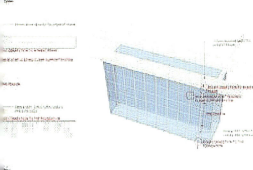

二等奖

藤本壮介建筑设计事务所和奥雅纳工程顾问公司（东京）
本页，停车场平面图
对页，竞赛展示板，显示了地形和景观设计

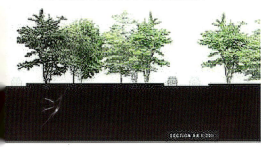

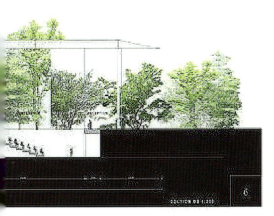

Centre for Promotion of Science of the Republic of Serbia

DIVERSITY

OUTDOOR SPACE - SCIENCE GARDEN

DAYLIGHT

120110sf

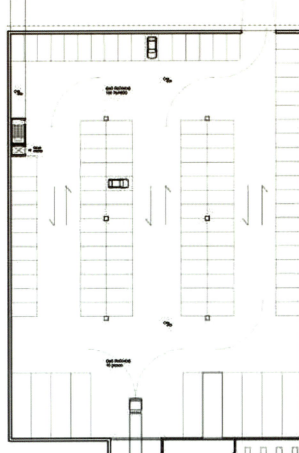

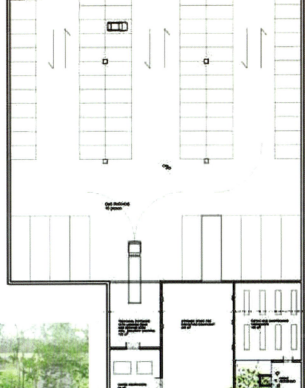

Second Prize

Sou Fujimoto Architects

with

Ove Arup
Tokyo

THIS PAGE
Floor plan, parking
OPPOSITE PAGE
Competition boards indicating the topo-graphical/landscape approach of the author.

ELEVATION SOUTH EAST

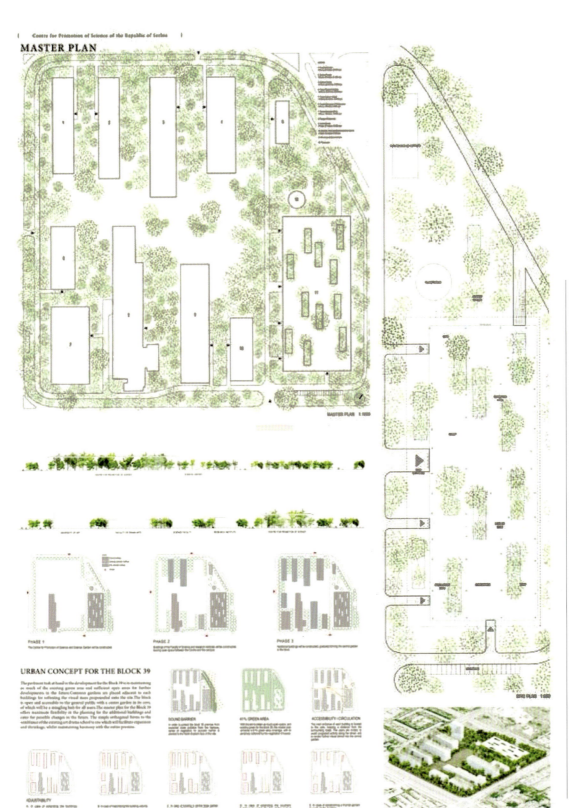

MASTER PLAN

Centre for Promotion of Science of the Republic of Serbia

URBAN CONCEPT FOR THE BLOCK 39

PHASE 1

PHASE 2

PHASE 3

120110sf

二等奖

藤本壮介建筑设计事务所和奥雅纳工程顾问公司（东京）

二等奖由来自日本的藤本壮介建筑设计事务所获得，与一等奖项目有许多相似之处。走进科学森林，建筑师突出了景观和造型结构元素，一丛丛的树木支撑了科学中心上方的华盖。一些功能区嵌入了景观之中，与华盖和树林融为一期。这一项目转变了自新贝尔格莱德建成以来我们对景观的理解和对自然统治的思想。

左图
景观设计图
对页
立面图

Second Prize

Sou Fujimoto Architects

with

Ove Arup Japan Pty. Ltd.
Tokyo

Second prize went to Sou Fujimoto Architects of Japan for a scheme that is similar in many ways. Entering the Forest of Science, the architects also highlight the land-scape and form structural elements around groups of trees to hold up a canopy above the program. Some of the program is then sunken into the landscape with this canopy and forest of trees towering above. This scheme suggests a change in our understanding of the landscape and our domination of nature since the regular grid of New Belgrade was first laid out. -TS

LEFT
Plan with landscape scheme
OPPOSITE PAGE
Elevations

Forest of Science

1. Vision
SCIENCE PARK

For the development of Block 39 and the Centre for Promotion of Science in the new Belgrade, we propose to create a Science Park. The Science Park is an open plateau connected to the city. It is a meeting place for all members of the community, to foster a dynamic and pedagogical environment for learning and engagement. The Science Park coloured rich in various activities undertaken by research community, supporting schools and citizens will contribute to the fruitful cultural and economic vitality of new Belgrade, and to the greater Serbia.

2. Features
FOREST in the city

Connecting with the peripheral green spaces, the development is equal to creating a network of green spaces, offering an affable asset to the city and its inhabitants. Akin to an oasis in the city, the places and its experience sets a tempo for the growth and life of a city which in turn affects its inhabitants.

BIG ROOF in the site

In order to facilitate the diverse spectrum of activities and the property of the park, a Big Roof is suspended at a height of 15 metres. The large continuous space added with the transparency below provides a welcoming access, as if one may just thereafter through and into a park. The horizontal continuant of the space accepts the activities and programs to spill in and out various flora of its users and inhabitants in the next.

GLASS PATIO in the architecture

By inserting layers of Glass Patios filled with trees within the building, one experiences a sense of being both inside and outside, feeling the exterior outside whilst being indoors. The new moments made by layers of reflection encourages an uncommented openness for discovery and interaction which in turn will lead to the greater dissemination of the Science and its culture into the City.

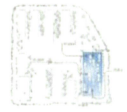 

3. Landmark
GREEN

This architecture is made from 2 types of Forests. One is the natural Forest of the site and the other is the Forest of information which permeate through the Science Centre. The cohesion of the 2 Forests is manifested in its appearance. As a Science Centre of the new century, the ephemeral characteristics of the natural greens and glass patios are pressed onto the ceiling plane 15m in the air. Pulsating with movement and time, the marriage of the 2 greens engenders a new identity for the Centre for Promotion of Science.

ARCHITECTURAL CONCEPT OF THE CENTRE FOR PROMOTION OF SCIENCE

The Centre for Promotion of Science in essence is one grand space covered by a big roof of 15 m ceiling height and divided by multiple glass patios encased with trees inside. The facade covering the building is made by transparent glass on all faces, visually open to the users, and the passerby. The program housed inside also shares this quality of being an open learning environment, flexible and approachable space, various rooms that prospectus seeks to foster a greater interest towards the discipline, and to bring the science closer to the general public, and moreover being a dynamic and engaging phenomenon for the future.

ZONING

The Science Centre comprises of four main areas, The Exhibition, Science-Club, Planetarium, and the Conference area.

The above areas are connected via the Lobby street, while being able to be partitioned and compartmentalized by movable partitions depending on its use, the entirety of the centre is also able to be opened. Service and supportive functions such as restaurants and cafe are also integrated to the above functions. The glass patios are loosely but thoroughly integrated within the Science centre, to provide segmentation at times, and at times visual vantage and leeway. One notable factor of the glass patios is its function to bring natural daylight and outdoor views into the deep inside of the premise.

Staff areas are located more independently, yet maintaining its connectivity with associative public programs. Other closed programs such as the laboratories are located in the basements, which contrary to conventional basement characteristics, is well lit through the light filled patios penetrating through the ground level.

ACCESS / CIRCULATION

The Science Centre has 1 main entrance and 3 sub-entrances located on the western face of the building. The Lobby street is capable of receiving numerous visitors, for which they can be directed to the different destinations of the centre. The Lobby street facilitates the information center and acts as a portal for admission to all public functions; logistic staff entrance is located at the southern side of the building, for a clear demarcation of the staff with the public, or at times, for exclusive attendees to conferences. The loading yard located in the basement is vertically connected to the Exhibition spaces in the case of loading exhibition items and heavy machinery.

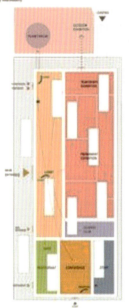

FUNCTIONAL ZONING PLAN

FLEXIBILITY

The zoning strategy implemented is a very loose one. The centre being capable of flexibly changing its size and shape according to the scale and characteristics of the activity. The exhibition area can be used as a grand singular space while also being able to be divided into a number of smaller areas by a mobile modular wall system. In case of a large-scale exhibition, the entire exhibition area and lobby space can be used, whereas only two or three rooms may be used for a smaller exhibition, leaving the rest of the exhibition area open to the public.

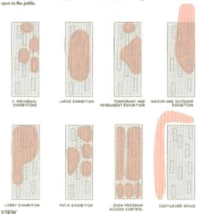

VIEW

The wall-free full transparency of the space achieves a 360-degree view, beyond through the greenery past the building and above to the sky.

Furthermore, this endows a clear navigation system being able to assess ones' location instantaneously.

120110sf

Second Prize

**Sou Fujimoto
Architects**

with

**Ove Arup Japan Pty.
Ltd.
Tokyo**

LEFT
Forest with canopy
OPPOSITE PAGE
Plan

二等奖

藤本壮介建筑设计事务所和奥雅
纳工程顾问公司（东京）

对页
森林和华盖
对页
规划图

TEMPORARY EXHIBITION AREA 2 500 m2

PERMANENT EXHIBITION AREA 2 400 m2

ENTRANCE HALL AND MAIN ENTRANCE 100 m2

MAIN ENTRANCE

PERMANENT EXHIBITION AREA1 600 m2

PERMANENT EXHIBITION AREA 4 500 m2

PERMANENT EXHIBITION AREA 3 500 m2

SCIENCE PLAYGROUND AREA 100 m2

CONFERENCE HALL

RESTAURANT 240 m2

PLAN 1 p. 6 00 DOOR

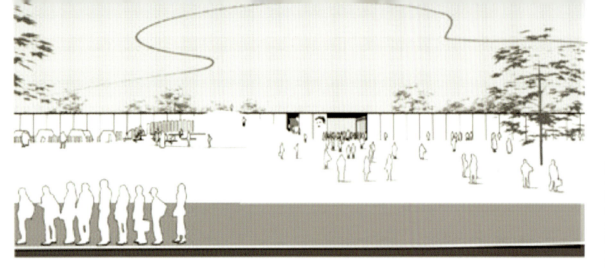

Third Place

ARCVS
Belgrade, Serbia
Team
Branislav Redzic´, Dragan Ivanovic´
Zoran Milovanovic´, Zoran Dorovic´
Vesna Milojevic´, Boris Husanovic´
Marko Todorovic´, Predrag Stefanovic´
Consultants
Sreto Kuzmanovic´, structural
Slavisa Milosavljevic´, traffic

LEFT
Pedestrian perspectives to entrance
BELOW
Birdseye views

OPPOSITE PAGE
Views of site from distances
View along exterior
Inside view

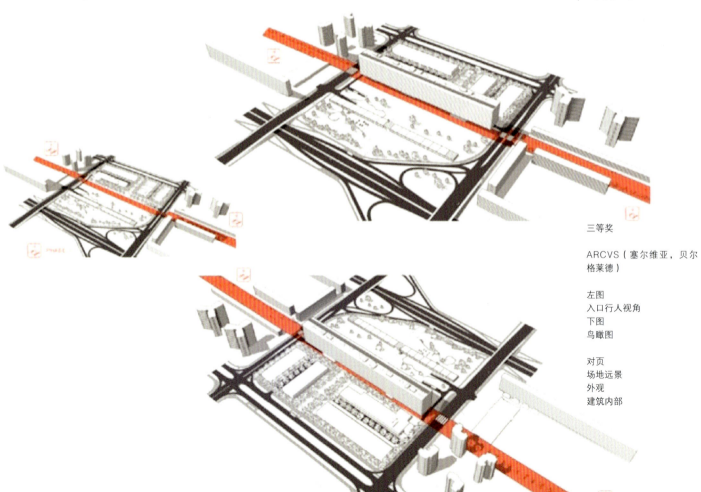

三等奖

ARCVS（塞尔维亚，贝尔
格莱德）

左图
入口行人视角
下图
鸟瞰图

对页
场地远景
外观
建筑内部

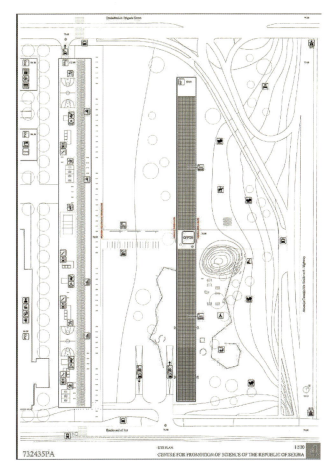

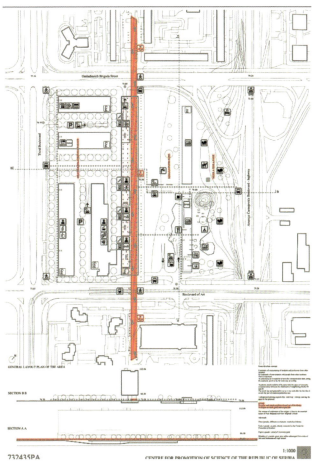

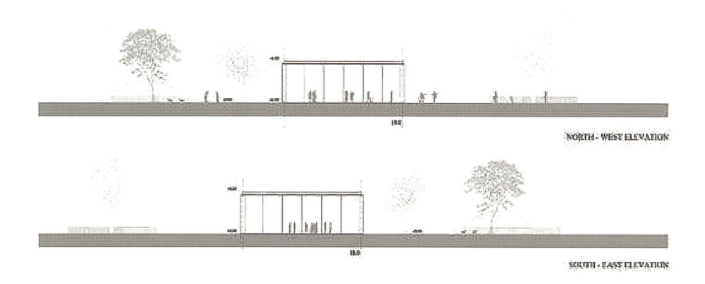

NORTH - WEST ELEVATION

SOUTH - EAST ELEVATION

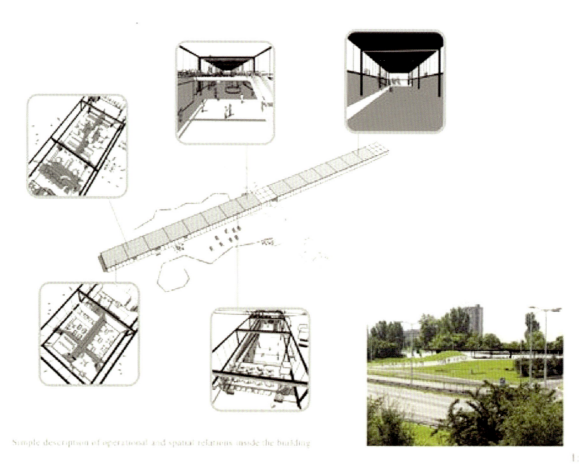

Simple description of operational and spatial relations inside the building

732435PA

1:200

CENTRE FOR PROMOTION OF SCIENCE OF THE REPUBLIC OF SERBIA

8

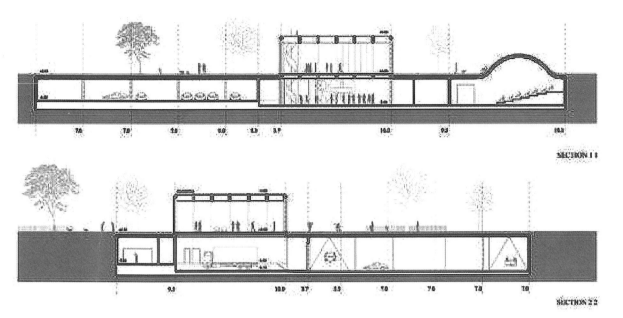

SECTION 1 1

SECTION 2 2

Third Place

ARCVS
Belgrade, Serbia

LEFT
Sections
BELOW
Plans

OPPOSITE PAGE, ABOVE
Elevations
OPPOSITE, BELOW
Perspective renderings
Aerial view of site

三等奖

ARCVS（塞尔维亚，贝尔格莱德）

左图
剖面图
下图
规划图

对页，上图
立面图
对页，下图
效果图
场地鸟瞰图

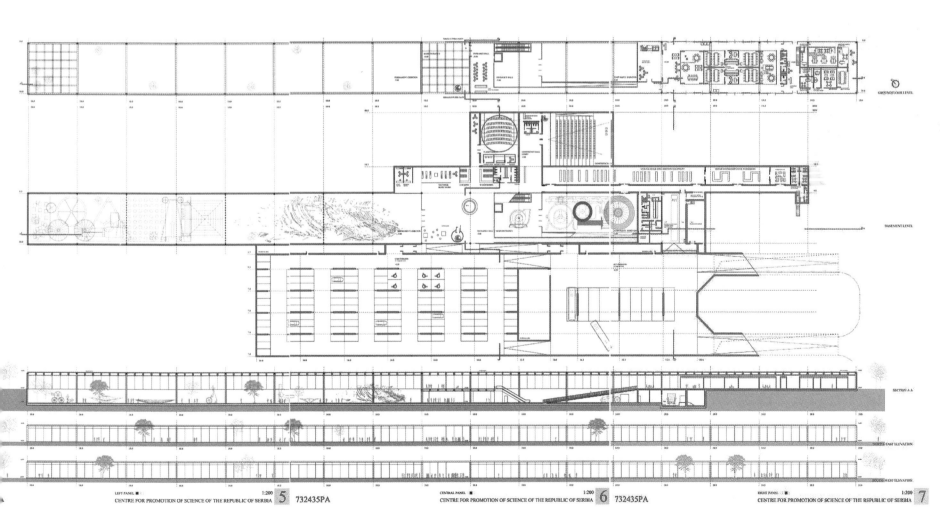

GROUND FLOOR LEVEL

BASEMENT LEVEL

SECTION A A

NORTH EAST ELEVATION

SOUTH EAST ELEVATION

LEFT PANEL ■   1:200   **5**   732435PA
CENTRE FOR PROMOTION OF SCIENCE OF THE REPUBLIC OF SERBIA

CENTRAL PANEL ■   1:200   **6**   732435PA
CENTRE FOR PROMOTION OF SCIENCE OF THE REPUBLIC OF SERBIA

RIGHT PANEL ■   1:200   **7**
CENTRE FOR PROMOTION OF SCIENCE OF THE REPUBLIC OF SERBIA

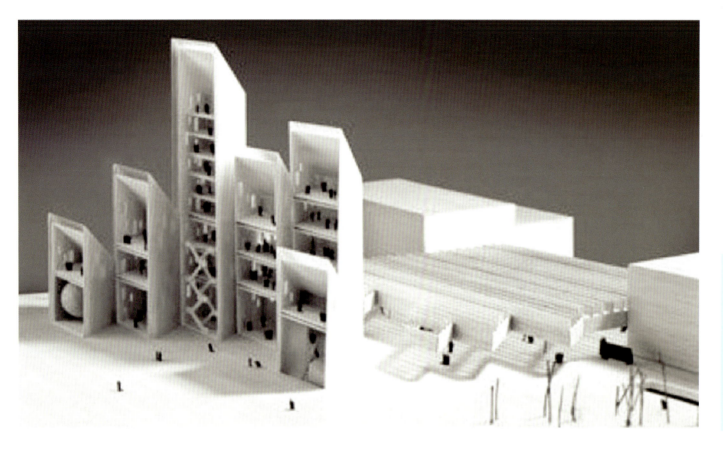

荣誉奖

佩斯克拉·乌拉吉建筑事
务所（西班牙，马德里）

左图
模型鸟瞰图
下图
效果图

Honorable Mention

Pesquera Ulargui Arquitectos, S.L.P
Madrid, Spain

LEFT
Birdseye view of model
BELOW
Renderings

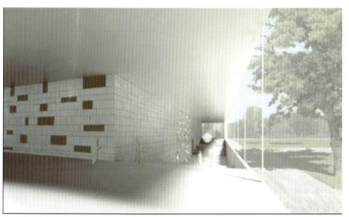

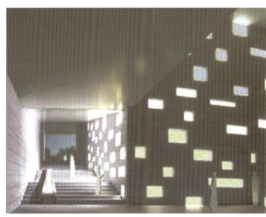

Honorable Mention

Durig AG, Jean-Pierre Durig
Zurich, Switzerland

ABOVE
View from boulevard
LEFT
Interior views
BELOW
Birdseye view of model

荣誉奖
杜里格AG；让-皮埃尔·杜里格
（瑞士，苏黎世）
上图
从林荫大道看项目
左图
建筑内部
下图
模型鸟瞰图

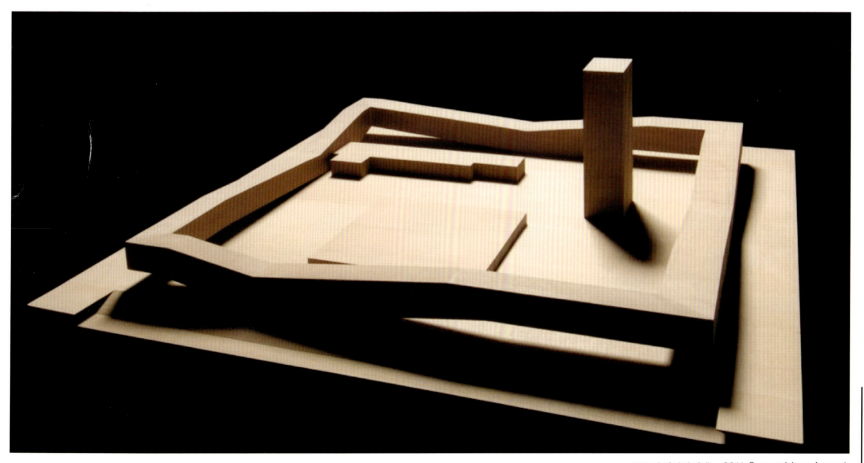

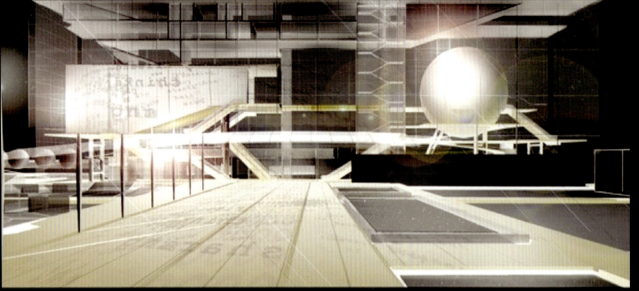

N ENTRANCE BY NIGHT

本页和对页
竞赛展示板，包括效果图、剖面图、平面图和模型鸟瞰图

Honorable Mention

Vladimir Lojanica
Belgrade, Serbia
Team
Sonja Pesterac,
Marija Corluka Mijovic
Vladimir Cvejic
Marija Konstandinidis
Nikola Ilic

THIS PAGE AND OPPOSITE
Competition boards with illustrations including renderings, sections, plans and birdseye view of model

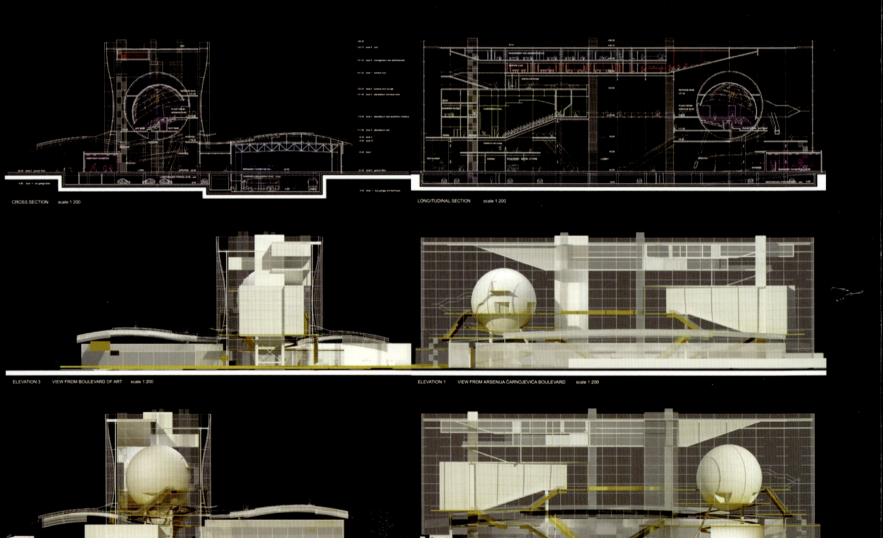

CROSS SECTION   scale 1:200

LONGITUDINAL SECTION   scale 1:200

ELEVATION 3   VIEW FROM BOULEVARD OF ART   scale 1:200

ELEVATION 1   VIEW FROM ARSENIJA ČARNOJEVIĆA BOULEVARD   scale 1:200

CROSS SECTION a-a   scale 1:1000

CROSS SECTION b-b   scale 1:1000

3D VIEWS OF THE LOCATION

Arsenija Carnojevica Boulevard - highway

Roof garden

Science Institute

Centre for promotion of science

Science garden

University Campus

Faculty of Drama Arts

future additional area   future additional area

GENERAL LAYOUT PLAN   scale 1:1000

The Barge 2011 Design Competition 2011驳船码头设计竞赛

Innovative Design Finds Anchorage in Boston 创意设计在波士顿生根发芽

by Dan Madryga

丹·马德里加

Fort Point Channel is poised to become Boston's "next great place." As a centrally located connection between the waterfronts of downtown and South Boston, the long underutilized waterway has become the subject of attention in the last decade. The award winning 2002 Fort Point Channel Watersheet Activation Plan has laid the groundwork for a revitalized waterfront neighborhood that promises a variety of recreational, cultural, and maritime facilities. This September, in addition to the usual signs of revitalization, visitors may be surprised to come across a bright pink balloon hovering over the channel. This intriguing object is the result of SHIFTboston's recent Barge 2011 Design Competition.

For the past three years SHIFTboston has been holding annual ideas competitions to generate and promote innovative design concepts that address issues in the urban environment. This year's focus on the Fort Point Channel is an attempt to further enliven what has long been a rather featureless body of water, and unlike the previous SHIFTboston competitions, the winning entry will actually be built and open to the public as a temporary exhibit.

水前区港道致力于成为波士顿的"第二个标志性景点"。人们于20世纪初开始注意这条连接了市中心水岸和南波士顿的未被充分利用的水道。2002水前区港道水岸活跃计划的获奖作品为复兴水岸区奠定了基础，预计打造一系列娱乐、文化和航运设施。2011年9月，除了普通的复兴设施之外，游客们还能惊喜地看到一个亮粉色的气球盘旋在港道之上。这个有趣的物体是城市规划组织——"改变波士顿"所举办的2011驳船码头设计竞赛的产物。

近三年以来，"改变波士顿"一直通过举办年度概念竞赛来产生和宣传创意城市规划设计理念。本年度他们聚焦于水前区港道，旨在活跃这个毫无特色的水体。与之前的竞赛不同，这次的优胜设计将被建成并作为临时展览。

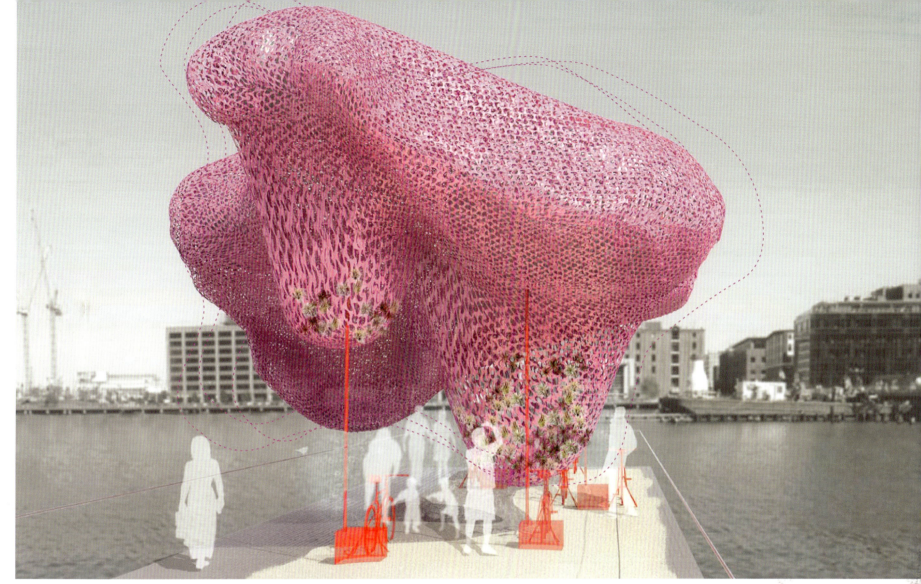

Pink Plastic Front Side

3D Digitally Fabricated Pattern

Reflective Mylar Back Side

Solid Areas Retain Less Water

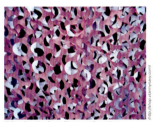
3D Digital Camouflage Distorts Silhouettes and Retains Water

First Place
"Lighter Than Air"
Rachely Rotem & Phu Hoang
New York City, NY

Opposite
View From Congress Street Bridge
Above
View of "Camovapor" Climate System from Seating Lounge

一等奖
"轻胜空气"
拉切利·罗特姆和黄普工作室（纽约）

对页
从国会街大桥看设计
上图
从休息区看"迷彩蒸汽"气候系统

The challenge of Barge 2011 was simple enough: to create a design which could be installed on a barge anchored off the channel. Architects, artists, and designers were invited to "unleash their wildest dreams," with a strong emphasis on experimentation and exploration. As such, the program requirements were kept rather open-ended to allow for a wide variety of inventive responses and solutions. Entrants were encouraged to develop design concepts that would create a unique sensory experience, using innovative sustainable materials and methods for a closed-loop electrical system that would not require grid-based energy. As the winning design is to actually be constructed, competitors were asked to keep within an estimated budget of $10,000, an amount that would be rewarded to the winning team to cover fabrication and construction costs.

In March, SHIFTboston received 102 entries from national and international participants. Twenty-one of these submissions were initially selected by an impressive multidisciplinary jury to move to the final round.

After narrowing the field down to six finalists. the jury settled on "Lighter Than Air" as the winning design.

2011驳船码头竞赛的挑战十分简单：打造一个能够安装在港道驳船码头上的设计。建筑师、艺术家和设计师都可以参与进来，"释放他们狂野的梦想"。因此，竞赛要求十分开放，允许参赛者提交形形色色的创意设计。竞赛鼓励参赛者利用创新的可持续材料和闭合回路电力系统来开发提供独特感官体验的设计理念。由于优胜设计将被真实建造，竞赛参与者必须将预算控制在10,000美元——这些钱将是优胜团队的奖金，用于支付建造成本。

三月，"改变波士顿"共收到了102件来自国内的参赛作品。其中的21件被评审委员会选出，进入了下一轮入围。在进一步缩小入围奖为6件作品之后，评审们将"轻胜空气"选为优胜设计。

Lighter than Air

Camouflage as Concept

A barge floating at dock in Boston's Fort Point Channel is a familiar sight from Congress Street Bridge above. The silhouette of the barge is recognizable from the industrial past of Boston's waterfront. "Lighter than Air" would create a climatic environment on this barge—temporarily transforming both the experience aboard as well as its visual presence in the channel. Principles of camouflage are used as a design strategy to create an atmospheric phenomenon, transfiguring the barge into an interactive sensory environment. Camouflage makes a familiar object appear strange—or an unfamiliar object is made to appear familiar. "Lighter than Air" uses camouflage as concept, though it is used to transform space instead of obfuscating space.

Transfiguring the "Camovapor"

"Lighter than Air" would produce an environmental phenomenon in Boston's Fort Point Channel. The project is comprised of two layers of a 3-D digitally fabricated camouflage net suspended by balloons, forming the "camovapor" climate system that floats above the barge deck. Weather balloons filled with either helium or hot steam would levitate the 3-D camouflage nets. The "camovapor" climate system would constantly change its shape and orientation, responding to both wind forces and interactions with the public. Views from below into this climate system would also change according to its mutable form. Four helium-filled balloons would provide enough lift to carry at least eighty pounds—more than enough to support the lightweight camouflage nets and air plants. The novel use of hot steam to inflate three additional weather balloons would initiate a hydrological cycle that converts water vapor to condensed distilled water. The hot steam balloons create an interactive experience for the public—lifting, lowering and transfiguring the "camovapor" unpredictably.

Sensing Micro-climates

"Lighter than Air" would not use any electricity or potable water from the city. An outdoor gym of bicycles provides "pedal power" to generate steam from captured harbor water. Visitors ride bikes to create energy; this energy is captured by a pedal generator and stored with a DC battery. The energy would power portable steam generators; the steam created would be used to either lift the weather balloons or to transport humidity directly into the "camovapor" climate system. As the hot steam condenses on the cooler surfaces of the balloons and camouflage nets, the resulting distilled water would trickle down the walls of the "camovapor" and irrigate the air plants. The 3-D perforation of the camouflage net would temporarily retain this water, creating cool micro-climate zones when combined with the harbor breeze. Eventually, the water would be drained beneath a stone mulch floor and returned to the harbor itself. "Lighter than Air" uses the principles of atmospheric micro-climates to create a new public space. These micro-climatic spaces, each with their own temperature and relative humidity, would provoke the public to interact with weather as an ephemeral form of architectural space.

Material After-life

The design process of "Lighter than Air" considered both for the "life" of the pavilion as well as its "after-life." It is important to plan for the lifecycle of materials when designing for a temporary event. All of the materials are locally sourced and have been considered for their recyclability. The 3-D camouflage nets are made of recycled plastic and would be donated to local schools after the event. The bicycles would be purchased from used bicycle shops. The "bike to steam" generator systems would also be donated to local schools. Lastly, the stone mulch floor material is made of recycled stone content.

1. Condensed Water Drains Below Floor to Harbor
2. Rock Mulch Thermal Floor
3. Bean Bag Lounge Seating
4. "Camovapor" Climate System
5. Tethering Line for "Camovapor"
6. Helium Balloons
7. Evaporative Cooling Micro-climate
8. Hanging Airplants
9. "Bike to Steam" Generator
10. Stationary Bicycles
11. Steam Pipe
12. Condensed Water Droplets
13. Hot Steam Balloon
14. Condensed Water Collected Inside Balloon
15. Wind Currents Moving Through "Camovapor"
16. Boundary Movements of "Camovapor"
17. Wind Currents Moving Around Balloons
18. Water Storage Tank
19. Water Inlet with Passive Water Ram Pump

1. Urban Pattern

2. Disruptive Camouflage Pattern

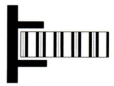

3. Site Boundary

4. Boundary Disruption

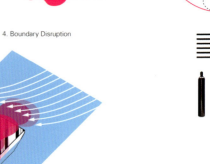

HELI

Portable Heli
- Tank can be reac
- 260 cubic feet pe
- Can refill a 10 fo
- Balloons hold he

Balloon Lift Diagram

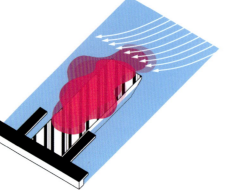

5. "Camovapor" Climate System Changes According to Wind and Public Interaction

Camouflage Concept Diagram

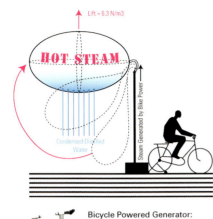

Lift = 6.3 N/m3

HOT STEAM

Condensed Distilled Water

Steam Generated by Bike Power

load of a 10' meter Helium oon is 12 pounds

east 20 times

Bicycle Powered Generator:
- Attaches to any bicycle
- 300 watts per generator
- www.econvergence.net/electro.htm

Portable Battery:
- Stores energy from generator for later use
- 400 watt battery
- www.econvergence.net/electro.htm

Portable Steam Generator:
- Creates steam from heating water
- Portable for outdoor use
- www.smartpakequine.com

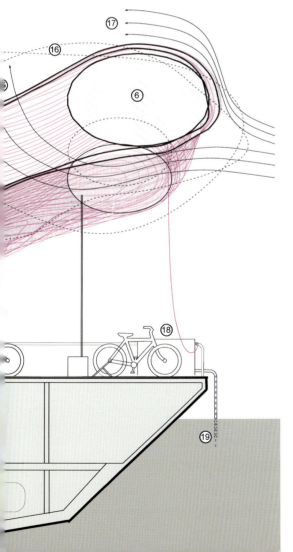

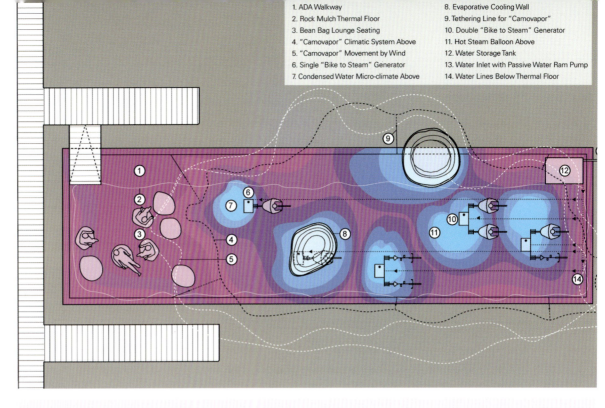

1. ADA Walkway
2. Rock Mulch Thermal Floor
3. Bean Bag Lounge Seating
4. "Camovapor" Climatic System Above
5. "Camovapor" Movement by Wind
6. Single "Bike to Steam" Generator
7. Condensed Water Micro-climate Above
8. Evaporative Cooling Wall
9. Tethering Line for "Camovapor"
10. Double "Bike to Steam" Generator
11. Hot Steam Balloon Above
12. Water Storage Tank
13. Water Inlet with Passive Water Ram Pump
14. Water Lines Below Thermal Floor

First Place

"Lighter Than Air"
Rachely Rotem
Phu Hoang
New York City, NY

"轻胜空气"
拉切利·罗特姆和黄普工作室（纽约）

"轻胜空气"由来自纽约的建筑师拉切利·罗特姆和黄普共同打造。他们利用平凡的材料进行了大胆的设计，制造了一个集合了创新和奇思妙想的装置。充满了氢气的聚酯薄膜气球被亮粉色的迷彩网包裹着，打造了一个球根状的结构，悬浮在驳船码头之上，随着微风和人的触碰改变着造型。从远处看，这个明亮的超现实装置为周边的环境注入了鲜活的色彩和能量。

在近距离观看，这个装置也极具吸引力。设计中，与驳船码头地面和气球相连接的固定式健身脚踏车独具特色。蹬踏脚踏车将激活一系列的气泵，为气球充气。脚踏车的额外能量将被储存在驳船甲板下方的电池中，帮助气球结构上浮和移动。除了具有互动价值之外，这些脚踏车还帮助展示了一个发电的创意方法。

"Lighter Than Air" is a collaboration between Rachely Rotem and Phu Hoang, New York based architects who also teach at the University of Pennsylvania. The duo proposed a bold design that uses surprisingly mundane materials to create an installation that combines innovation with whimsy. Mylar balloons filled with helium and wrapped in bright pink camouflage netting create a bulbous structure that will hover above the barge and subtly change its shape depending on the wind and human interaction. From afar, this bright, surreal object will interject some much-needed color and liveliness into its surroundings.

The installation will be equally intriguing at close range. The design features a unique interactive element in the form of stationary exercise bikes situated on the floor of the barge and connected to the balloons. Peddling the bikes will generate energy to activate a series of pumps that will push air into the balloons. Excess energy from the bikes will be stored in batteries below the deck of the barge, which will help keep the balloon structure buoyant and moving in times of low usage. Aside from providing an interactive activity, the bikes will help illustrate an innovative method of generating off-the-grid power.

In the coming months, Rachely Rotem and Phu Hoang's winning design will be undergoing the necessary design modifications and refinements necessary to turn their innovative and intriguing vision into reality. With such a thoroughly conceived concept, there is little doubt that the finished product will be remarkably faithful to the initial idea. The "Lighter Than Air" barge will be open to visitors during September and October of 2011, with an opening event planned for September 16th.

While "Lighter Than Air" is certainly worthy of first prize, the other five finalist projects illustrate the impressive range of innovative and thought-provoking ideas that were born out of the Barge 2011 design competition. Balancing aesthetics, invention, and practicality, almost any of these shortlisted designs would have made for interesting built pieces. -DM

ABOVE AND LEFT 上图和左图
Competition board 竞赛展示板

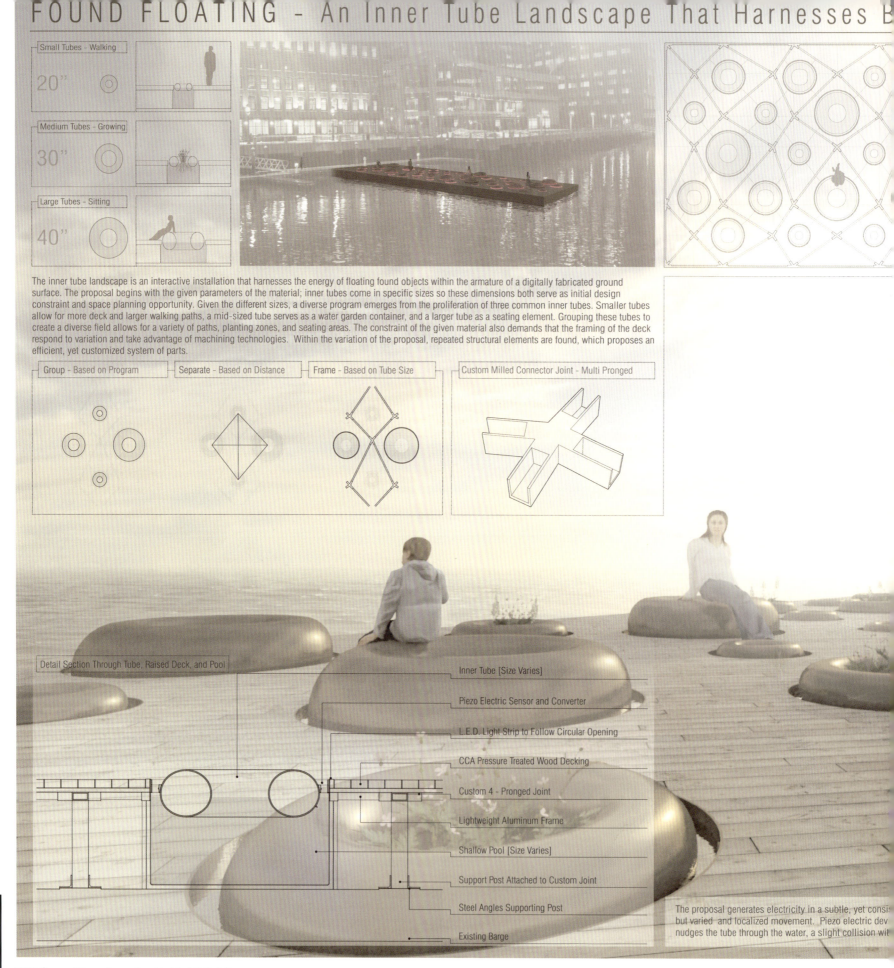

Small Tubes - Walking
20"

Medium Tubes - Growing
30"

Large Tubes - Sitting
40"

The inner tube landscape is an interactive installation that harnesses the energy of floating found objects within the armature of a digitally fabricated ground surface. The proposal begins with the given parameters of the material; inner tubes come in specific sizes so these dimensions both serve as initial design constraint and space planning opportunity. Given the different sizes, a diverse program emerges from the proliferation of three common inner tubes. Smaller tubes allow for more deck and larger walking paths, a mid-sized tube serves as a water garden container, and a larger tube as a seating element. Grouping these tubes to create a diverse field allows for a variety of paths, planting zones, and seating areas. The constraint of the given material also demands that the framing of the deck respond to variation and take advantage of machining technologies. Within the variation of the proposal, repeated structural elements are found, which proposes an efficient, yet customized system of parts.

Group - Based on Program

Separate - Based on Distance

Frame - Based on Tube Size

Custom Milled Connector Joint - Multi Pronged

Detail Section Through Tube, Raised Deck, and Pool

Inner Tube [Size Varies]

Piezo Electric Sensor and Converter

L.E.D. Light Strip to Follow Circular Opening

CCA Pressure Treated Wood Decking

Custom 4 - Pronged Joint

Lightweight Aluminum Frame

Shallow Pool [Size Varies]

Support Post Attached to Custom Joint

Steel Angles Supporting Post

Existing Barge

The proposal generates electricity in a subtle, yet consi
but varied and localized movement. Piezo electric dev
nudges the tube through the water, a slight collision wit

2011 Barge 2011 Design Competition
2011 驳船码头设计竞赛

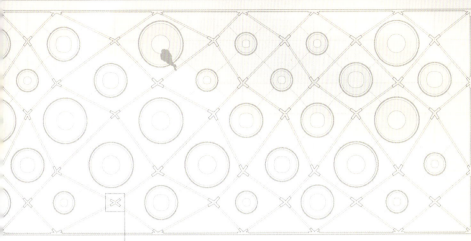

入围奖

"创立悬浮"
安东尼·迪·马利（波士顿）

"创立悬浮"是入围作品中最简约的设计，依赖微妙的特征来提供单一体验。设计以嵌入不同尺寸的内胎的平台为基础——较大的轮胎适合坐下休息，而较小的则可用作花盆。这些内胎的运动与行人形成了互动。水流的不断运动将通过压电现象产生能量，被储存起来，用来点亮浅紫色的灯泡，为驳船码头营造了迷人的夜景。由于整体设计理念具有创新特征，它低调的特性或许不易于吸引公众的好奇心。——丹·马德里加

e proximity of the pool's edge to the tube, the movement of the water, and the resiliency of the rubber allows for a constant, kinetic energy into electricity through passive means. These devices are affixed to the outer edge of the tube. As water triggers the device and powers an L.E.D. lighting strip.

Finalist

"Found Floating"
Anthony Di Mari
Boston, MA

"Found Floating" is the most minimalist of the finalists, relying on subtle features to provide a singular experience. The design is essentially a flat platform inset with various size innertubes — the larger tubes suitable for seating and the smaller ones functioning as plant holders. The movement of the innertubes, created by passenger interaction as well as the constant motion of the water, generates energy through piezoelectricity, which would be harvested and used to illuminate faint purple lights, giving the barge an intriguing nocturnal glow. While the overall concept is original and innovative, its understated nature might have had difficulty attracting curiosity or interest among the public. Its minimalism borders on the mundane, and at least one judge lamented the decision to cover up the design's interesting diamond shaped support structure with a rather unremarkable wood plank floor. -DM

LEFT
Competition board
竞赛展示板

Finalist

"Capullo!"
Jason Ejzenbart
Kara Burman
University of Manitoba
Winnipeg, Canada

入围奖
"卡普洛！"
詹森·埃森巴尔特（曼尼托巴大学）（加拿大，温尼伯）

"Capullo!" features an organic, translucent structure made of fiberglass rebar and industrial strength plastic wrap that would create a series of cave-like interior spaces. The design is one of the most responsive to exterior conditions. In rainy weather, rainwater would filter off the walls and into funnels, creating vaguely musical sounds. Meanwhile, a grated floor would allow visitors to see the rainwater collecting below.

Nighttime would offer another unique experience thanks to LED rope lights that would line the exterior skin, giving the translucent form a diffuse glow. The jury was most receptive to the introduction of a distinct, lighted form into a body of water that is normally rather dark and featureless at night -DM

RIGHT
Competition board

"卡普洛！"的透明有机结构由纤维玻璃条和工业强度塑料膜组成，制作了一系列的洞穴式室内空间。设计对外部条件反应十分敏感。在雨天，雨水被滤出墙壁，进入漏斗，发出依稀的音乐声。同时，格栅地面让游客们可以看到下方的雨水收集。夜晚，装置外表面的LED灯将打造独特的体验，为透明的造型增添弥漫的光线。评审们认为在夜晚的光亮造型远比黑暗而毫无特色的装置更受欢迎。——丹·马德里加

右图
竞赛展示板

Concept:

Beacon of glowing light seen from the surrounding area. The sound of raindrops landing on the canopy. Water rushing under your feet. An ever-changing topography moving through. The glow of the city lights softened around you. Breaks of fresh, salty air. Purposely highlighted views. Wind whistling around you. Large open spaces. Small tighter spaces. A warm, comforting cocoon.

This space is created to interact with rain, wind and natural and artificial light. It creates a visual and aural experience that accentuates these conditions. Simple materials are chosen that highlight the experience for the observer.

Rainwater landing on the canopy and rushing down through the funnels and onto the tin underneath will each have its own unique sound, creating a symphony of different instruments. There is an open funnel in the center of the installation where rainwater can fall in and be felt but also can be a vent for hot air to escape. The installation allows for a rippling effect in the plastic canopy and vibrations in the flexible rebar structure. As the space is walked through, the spaces undulate smaller and bigger, higher and lower ceilings. The lighting from inside the space allows the installation to glow in the evening. The waves on the surface of the channel's water contribute to refraction of light on the outside of the plastic covering. Natural light filters through the skin during the day and the colourful lights of the city to glow through into the space during the evening.

Materials:

The materials chosen to create this space are fiberglass rebar, conventional rebar, industrial plastic wrap, tin flashing and metal grating. Tin flashing is set on a gradient over the barge surface to allow rainwater to flow off the barge. The level metal grating set above this is to be walked on allowing the observer to see the rainwater under their feet. Conventional rebar is constructed in a grid pattern which creates seating areas and works structurally to keep the metal grating level and accessible. Because of the flexibility of fiberglass rebar, it is used to create arches for the canopy. It is clamped on both ends to the metal grating. Using different lengths of rebar and setting them in groupings, bending them in different directions towards the edge of the barge creates a varying topography. To enclose the space industrial plastic wrap weaves in and out of the fiberglass rebar skeleton to create a waterproof skin. A benefit of the flexibility of the material is that it can be altered easily to create many different forms that can open or restrict the space to change its function. The space can evolve from a general public space to a gallery or exhibition space or adjust to any function that the users see fit.

Energy:

Night lighting is incorporated in two separate ways to provide both stationary and moveable lighting options. Rope LEDs are intertwined in to the outer skin situated in the funnels. These are charged by small solar panels that are attached to the top of the canopy. Then small portable solar LED lamps that are often found in gardens are placed throughout set in the grating which visitors can move as they see fit to create there own lighting conditions.

Environment:

Boston has an abundance of used construction materials that are sold at reasonable prices; this is where the conventional rebar, tin flashing and metal grating would be purchased. Since the rebar of both kinds is barely altered, any construction company can reuse it.
The plastic wrap used as the skin is intended as a statement of the misuse of unrecyclable products. Plastic wrap along with other polymer products are often very difficult to recycle or reuse, it is because of the strength and convenience that they are used. This installation is a statement that demonstrates this wastefulness. It is done to let the public recognize how wasteful it is to use a material like this.

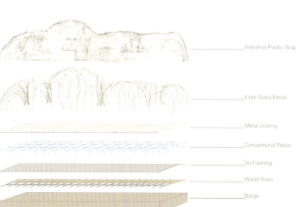

Industrial Plastic Wrap

Fiber Glass Rebar

Metal Grating

Conventional Rebar

Tin Flashing

Wood Truss

Barge

The facade flexes with the touch of a person or a breeze from the wind.

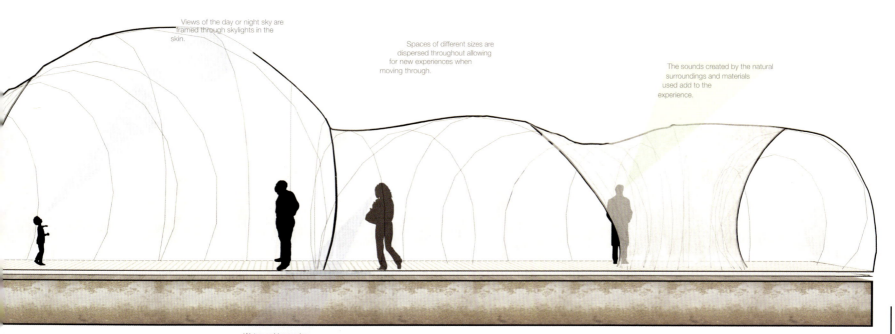

Views of the day or night sky are framed through skylights in the skin.

Spaces of different sizes are dispersed throughout allowing for new experiences when moving through.

The sounds created by the natural surroundings and materials used add to the experience.

Water rushing under observer's feet creates a visual and aural affect.

growboston.

Grow Boston proposes a two month long event in which the community works together towards the creation of a garden that slips into the Fort Point Channel waters as it outgrows the surface of the barge. As a play on the city's history of claiming land from the sea as it continually expands, the GROW Boston garden will increase its impact on the water front over its two month life span as floating satellite gardens slide off of an undulating landscape apparatus constructed on the barge surface. The apparatus is constructed from leased scaffolding, reclaimed plywood paneling, and recycled paper tubes. The apparatus is skinned with chicken wire gabion cages holding reused plastic bottles that are topped with donated plants. These floating gabion gardens are constructed by the installations visitors. The cages themselves are preassembled and stored on the barge, but the plastic bottle fill and plant material that sits on them is collected and donated by the visitors themselves. As more people visit the installation the over the first few weeks the barge's surface will be covered with small modular gardens. And as gardens continue to be added to the apparatus they will begin to spill out into Channel's water way creating a new artificial garden landscape.

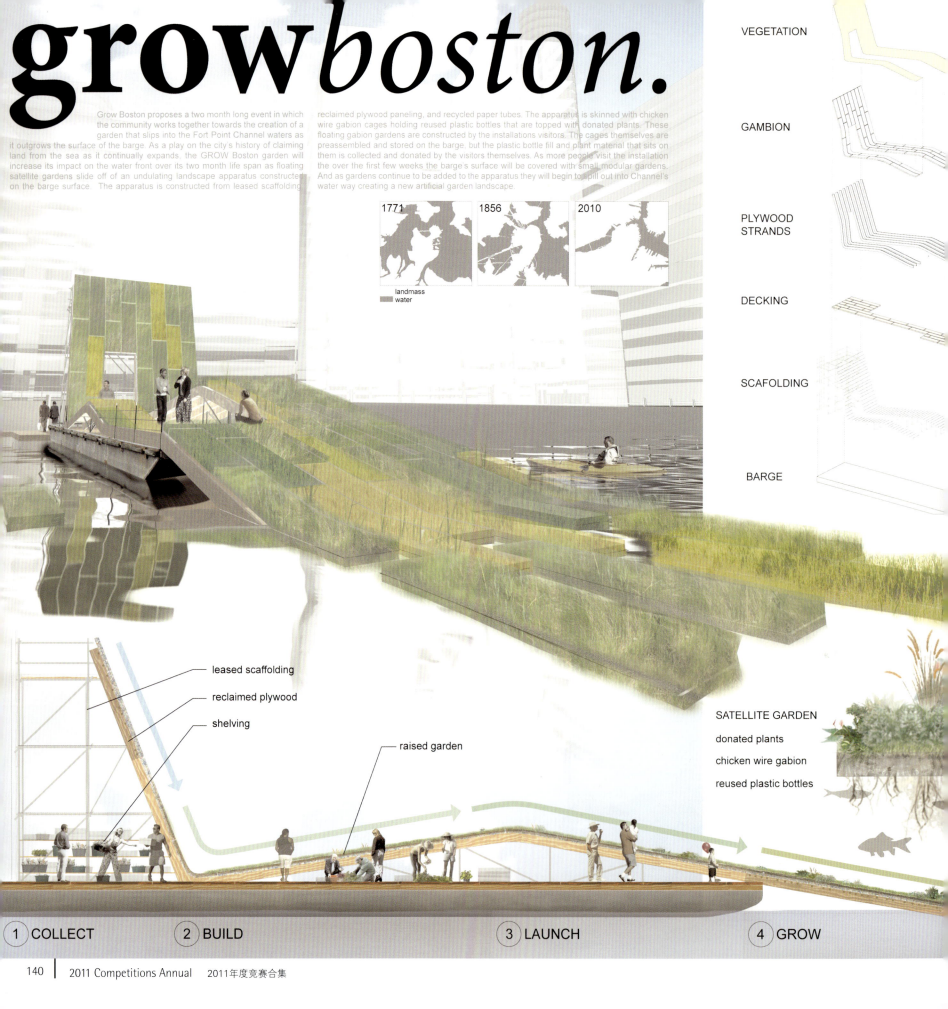

1771   1856   2010

landmass
water

VEGETATION

GAMBION

PLYWOOD STRANDS

DECKING

SCAFOLDING

BARGE

leased scaffolding

reclaimed plywood

shelving

raised garden

SATELLITE GARDEN

donated plants

chicken wire gabion

reused plastic bottles

(1) COLLECT   (2) BUILD   (3) LAUNCH   (4) GROW

Finalist

"Grow Boston"
Blake Thomas
Peter Broeder
University of Kansas
Lawrence, KS

"Grow Boston" by Blake Thomas and Peter Broeder of the University of Kansas envisioned the barge as a floating garden inspired by Boston's history of acquiring land by reclaiming the sea. Small block units constructed of chicken wire would provide self contained planters that, when joined together, appear to be a whole garden that gradually slopes down to the water. The design encourages public interaction and alteration by allowing visitors to bring their own plants to add to the garden. As more garden units are added to the top of the garden, the bottom blocks would be pushed off and proceed to float off into the Channel. The concept is certainly compelling, but one might wonder how effective and manageable it would be in practice. *-DM*

THIS PAGE AND OPPOSITE 本页和对页
Competition board 竞赛展示板

入围奖

"成长波士顿"
布莱克·托马斯；彼得·布洛德（堪萨斯大学）（劳伦斯市）

布莱克·托马斯和彼得·布洛德设计的"成长波士顿"从波士顿的填海造田中获得了灵感，将驳船码头看作一个悬浮花园。由细铁丝网制作而成的小块街区上有独立的花坛。当它们聚集在一起，便成为了一个缓缓滑入水面的花园。设计鼓励公众互动，允许人们在花园中种植自己带来的植物。当越来越多的花园元件被加入花园上方时，下方的街区便被推开，进入港道。这个概念十分引人注目，但是它的可行性值得讨论。——丹·马德里加

> COMMUNITY INVITED TO BUILD GARDENS

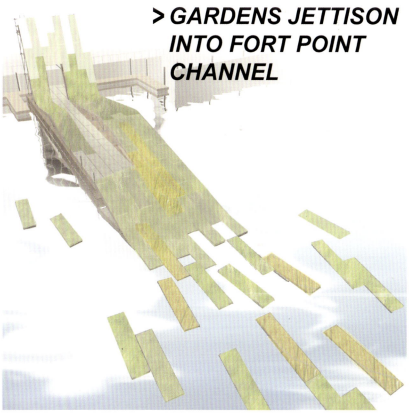

> GARDENS JETTISON INTO FORT POINT CHANNEL

Design Proposal:

Inspired by the water flow, the design was gradually staged to achieve the brief requirements as well as generating a topographic landscape feature on the waterfront. Harmonising water with structure, the water flow highlights the natural undulated curvature. The surface is a self-supporting structure; therefore there is no need for a secondary structure, leaving the shell elegantly thin.

We propose a green and economical way of on site construction by using prefabricated recycled materials. The shell surface is lightweight and easy to assemble on site, with thickness of the proposed plywood sheet is between 5-8 mm, while recycled reinforced steel ribs come in 4 mm thickness.
Using minimum material for a maximum undulating public open space.

Several circulation paths combined together, such as recycled water circulation, visitors' circulation and light circulation, define the surface's form.
The proposal is a timber-sculpted water garden that acts as a landscape feature on the edge of the water.

The water garden will benefit from an open space sitting spaces, a café/bar at the end of the barge for people to enjoy a drink while watching the sunset. Visitors will also benefit from water, light and greenery exhibitions, accompanied by the sound of water moving in and around the promenade.

Points of interest such as light reflecting ponds and elevated-ocean viewpoints ensure an extraordinary floating sensory.

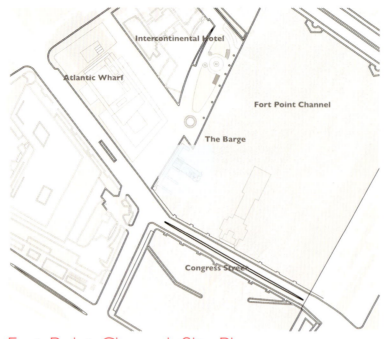

Fort Point Channel_Site Plan

入围奖

"淹没它！折叠它！"
苏慧·金；塔尔亚·桑丹克（韩国，法定洞）

"淹没它！折叠它！"的设计灵感来自于水的运动，设计师通过预制的回收材料打造了一个流畅的有机建筑结构。与其他参赛作品相反，设计师选择为驳船码头指派特定的功能，如水上花园、咖啡厅、酒吧和休息区。作为简单的建筑作品，设计缺乏典型的理念和独一无二的创新。同样，设计的整体审美价值也需要提升。

Barge Plan View

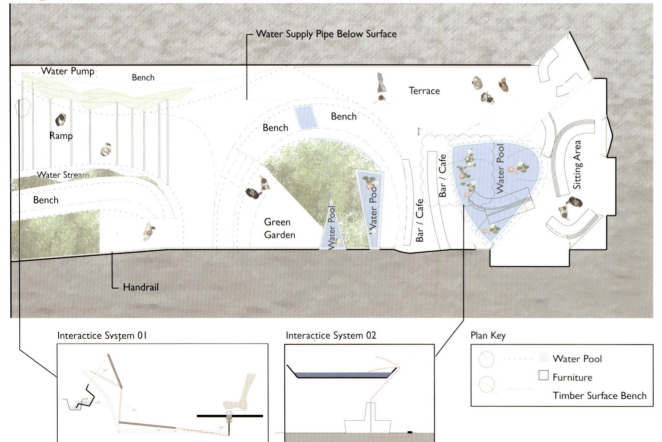

Finalist

"Flow it! Fold it!"
Soohyun Jin
Talya Sandank
Munbae-dong, South Korea

"Flow it! Fold it!" by Soohyun Jin and Talya Sandank of South Korea took its inspiration from the movement of water in creating a fluid, organic architectural form constructed out of prefabricated, recycled materials. In contrast to many of the other entries, the designers chose to assign very specific functions to the barge, including a water garden, café and bar, and sitting area. One of the more purely architectural submissions, the design nevertheless suffers from a lack of a clear defining concept or unique innovation. Likewise, the overall aesthetic could use some further refinement.

THIS PAGE AND OPPOSITE
Competition board
本页和对页
竞赛展示板

URBANQUIPU | (TALKING STRINGS)
creating an experiential catalogue of urban activity

SPATIAL PROCESION ELEVATION
scale 1"=10'

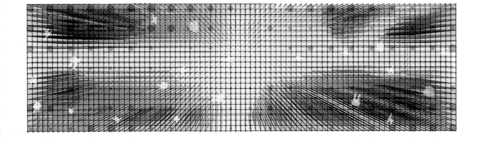

SPATIAL PROCESION PLAN
scale 1"=10'

THE QUIPU

(sometimes called talking strings) were recording devices used ancient civilizations. A quipu usually consists of colored, spun, and plied thread or strings from llama or alpaca hair. It was also often made of cotton cords. These civilizations would use these Quipus for relaying information, logging arithmetic, catalogueing events, and teling stories. They would use groupings of knots, varieties of knots and colors to convey this information.

URBAN QUIPUS

The urban quipus seeks a rediscovery of this method of communication through an urban instllation comprised of one grand quipu covering the barge carved by the activities that occur in a urban space and to be carved by the eventual activities that occur in it.

RECONCEIVED MATERIALS

The Urban Quipu will be framed from the edges of of the barge through salvaged ships masts. The above hanging grid will be woven and knot salvaged ship rope, and knotted again by the hanging strings of the Urban Quipu(also to be comprised of salvaged shiprope)

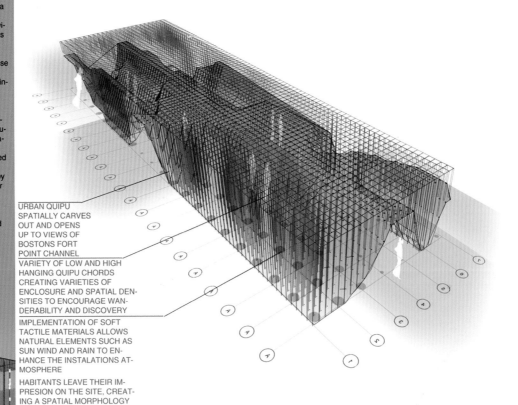

URBAN QUIPU SPATIALLY CARVES OUT AND OPENS UP TO VIEWS OF BOSTONS FORT POINT CHANNEL

VARIETY OF LOW AND HIGH HANGING QUIPU CHORDS CREATING VARIETIES OF ENCLOSURE AND SPATIAL DENSITIES TO ENCOURAGE WANDERABILITY AND DISCOVERY

IMPLEMENTATION OF SOFT TACTILE MATERIALS ALLOWS NATURAL ELEMENTS SUCH AS SUN WIND AND RAIN TO ENHANCE THE INSTALATIONS ATMOSPHERE

HABITANTS LEAVE THEIR IMPRESION ON THE SITE, CREATING A SPATIAL MORPHOLOGY THAT CATALOGUES THE EVENTS IN THE SPACE

THE URBAN QUIPUS WILL USE USE HYDRAULIC RAM PUMP TO FILL THE SURFACE OF THE BARGE WITH A THIN LAYER OF WATER TO REFLECT THE SKY AND THE HANGING ROPES TO CREATE A SERENE EPHEMEREL SPACE WITHOUT USING ANY OUTSIDE ENERGY

he barge I was trying to create a space
and manipulated by the activities actions
ned to happen there in. A space where
act not where they are. A space fo-
on the layering of activity, interaction (of
and demographics), physical discovery
otion. An installation focused on experi-

ting the installation as a machine for
welling. A space that encourages all
s(sitting, wandering, laying, watching the
l, children playing) observing and dia-
he activities I saw taking place in the
I diagrammed how children play from a
erspective. of children interact and to
e same sort of synergies existed. I then
ded to diagram how adults use a public
 park type space. Focusing on how they
ct/wander in these types of spaces.
n overlayed the diagrams deriving nodes
non circulations that would begin to ma-
 people in the space; as well as the
manipulating the space itself through
ng strings

Finalist

"Talking Strings"
Jason Ross
University of Tampa

One of the more poetic designs,
"Quipus – Talking Strings" takes
its name and design inspiration
from the ancient method of
recording important information
and events through the use of
multicolored strings knotted in
specific patterns to represent
words or meanings. In this design,
ropes are hung from a suspended
overhead structure that incorporates a salvaged ship. Hung at
varying heights, the ropes would
sway through contact with wind
and human interaction, providing
a semi-enclosed space that would
allow for partial views of the city
beyond. The floor of the barge
would be covered by a thin layer
of water acting a mirror, adding
an element of the surreal to this
intriguing space.

THIS PAGE AND OPPOSITE
Competition boards

入围奖

"会说话的绳子"
詹森·罗斯（塔帕布大学）

作为最具诗意的作品之一，"结绳语——
会说话的绳子"从古老的记事方法（古时
候的人们通过在不同色彩的绳子上打特别
的结来记事）获得了灵感。在这个设计
中，绳索被悬挂在一个打捞船的高架结构
上。不同高度的绳索随着风或者人的触碰
而飘动，打造了一个半封闭空间，可以远
眺城市的景色。驳船码头的地面上覆盖着
一层薄薄的水，增添了这个空间的超现实
氛围。

本页和对页
竞赛展示板

IMPROVING EDUCATION BY DESIGN

通过设计促进教育发展

Can reducing class size and hiring more qualified teachers be the only answer to improving our educational system? Not according to the goals set forth by the organizers of the most recent 2011 Cleveland Design Competition. This open contest challenged architects to come up with new ideas, keeping pace with "the continuing advancements in pedagogy, curricula and organization models." In other words, why shouldn't equal attention be devoted to the "reinvention of learning environments."

The subject of this competition was a new school in downtown Cleveland, the Campus International School, a public school featuring an international baccalaureate curricula. Although designed primarily for foreign students, it is also available to students residing in Cleveland and its suburbs. Based on the student body and curricula of the school, it would hardly qualify as your typical inner-city school. Still, the organizers believed that the competition would generate ideas that could be used not only in Cleveland, but also in other school districts throughout the country.

Of course, this is not the first competition with high expectations surrounding school design. As well-meaning as many of these competitions may have been, advances in that area have often been stymied by school facilities planners. Too often they find a way to water down progressive ideas, basing their objections on some obscure interpretation of general guidelines. One such example was the 2000 Chicago "Big Shoulders, Small Schools" competition won by Marble/Fairbanks. The winner of that competition was subjected by the school bureaucrats to unusual scrutiny, including the overly strict imposition of limitations concerning the ramping features of their design. The design was never built. Most disconcerting was that this all occurred while other, less innovative schools were being built throughout the city of Chicago. More lately, the Chicago school authorities seemed to have relented somewhat, as some projects by John Ronan have been able to survive the approval process.

Due to rising energy costs, sustainability has become a high priority in new construction for many school administrations. In San Francisco, for instance, some recent schools using solar energy have been able to attain a target of net zero energy consumption.

Besides looking for new models for learning within a physical context, the competition also had a strong social component. There is a common belief among Clevelanders that the future success of the city is dependent in large part on the success of its schools. Therefore, architects were asked explore how the design of better learning environments, and their ability to connect with the city, might help to slow or reverse the population exodus from the public school system in Cleveland, and thus many other urban areas throughout the world.

Once the adjudication process by the jury had taken place, it was obvious that the panel took the experimental nature of this process very much to heart. In addition to some very buildable projects, the jury singled out some for their purely provocative approaches. Also, it was clear that the entire project would have to be built in phases, for budget considerations, if for no other reason. So a totally integrated design, as good as it might be, would hardly be useful in these practical terms. But despite the high number of entries, and the quality displayed by some of the premiated designs, none apparently rose to the level to gain the kind of support one would need for serious consideration as a real project.

Jury
Kevin Daly, AIA
Design Principal-in-Charge, Daly Genik
Steven Turckes, AIA, REFP, LEED AP
K-12 Education Global Market Leader, Perkins + Will
David Mark Riz, AIA, LEED AP
Principal, KieranTimberlake
Amy Green Deines, Associate AIA
Associate Professor of Architecture, University of Detroit Mercy
Edward Schmittgen
Executive Director of Capital Planning and University Architect, CSU
Dr. Linda J. Williams Ph.D.
Superintendent, Various School Systems
Senior Director of Educational Services, WVIZ/Ideastream

缩小班级规模和聘用更优秀的教师是促进教育发展的唯一方案？2011克利夫兰设计竞赛的组织者说"不！"这个开放式竞赛要求建筑师提出全新的概念，与"教育学、课程体系和组织模型的持续发展"同步。换句话说，"改造教育环境"对促进教育发展同样重要。

本次竞赛的目标是在克利夫兰市中心建造一座新学校——国际学校，一所提供国际学士学位课程的公立学校。学校主要招收外国学生，也接受克利夫兰与其周边地区的学生。设计以学生主体和课程设置为基础，很难与典型的市内学校相同。然而，组织者坚信竞赛将产生全新的设计理念。这些理念不仅能应用在克利夫兰，还能应用在美国的其他地区。

当然，这并非是第一个高期待度学校设计竞赛。这些竞赛都充满善意，然而却总是受到学校规划者的阻碍。他们总是淡化先进的理念，制定含糊的规则。例如，由玛贝尔/费尔班克斯获胜的2000年芝加哥"大担当，小学校"竞赛。竞赛的优胜设计遭到了学校领导的反对，他们不认同设计的创新特色。这个设计一直没有得到实施。更让人沮丧的是，其他平淡无奇的学校设计反而在芝加哥城盛行。最近，芝加哥学校委员会似乎变得宽容了许多，约翰·罗南的一些项目已经获得了批准建造。日渐增长的能源成本让可持续性成为了新学校建筑的首要考量。在旧金山，一些学校利用太阳能来实现零能源消耗。

除了寻找新的学校建筑模型之外，竞赛还拥有强烈的社会责任感。克利夫兰人普遍认为城市未来的成功在很大部分取决于学校的成功。因此，建筑师必须探索如何设计更好的学习环境，学校与城市的连接将帮助缓和或逆转大批学生脱离克利夫兰（乃至全世界其他区域）公立学校的现状。

在评审裁决时，评审委员会十分注重设计的实验价值。除了一些可行的项目之外，评审还挑出了一些具有激进想法的设计。此外，由于预算原因，整个项目必须分期实施。因此，一个完整的设计在应用阶段很难起作用。尽管有诸多的参赛作品，获奖设计也品质优越，但还没有设计能够获得支持，获得建成的机会。

A NEW SCHOOL VISION

The **2011 Cleveland Design Competition: A New School Vision** invites professionals, students, firms and designers from all over the world to submit visions for a new K-12 public school in Downtown Cleveland.

At a time when educators are implementing dramatic new ideas in pedagogy, curricula and organization models, the reinvention of learning environments deserves equal attention. The 2011 Cleveland Design Competition presents an opportunity to re-imagine the school and explore how educational facilities must evolve to provide world class opportunities for learning.

cleveland design
competition

First Place

Michael Dickson
Brisbane, Australia

This design had to be considered a long shot, simply because the total composition of the plan included several multi-story structures-a marked departure from most new schools—unless they were in high-density metropolitan areas. But the jury evidently considered the positive aspect of this plan to be the superior integration of the various components as part of the whole.

On the philosophical side, some of the ideas proposed by the author would seem to be more appropriate to a prototype intercity public school, rather than one engaged in focusing on an extremely high level of traditional academic performance, as is the case with a baccalaureate program. Flexibility was one of the major aspects of this scheme, with a two-skin structure allowing for maximum changes, not just in the rearrangement of furniture, but also structurally to accommodate shifting priorities. -SC

THIS PAGE AND OPPOSITE
Competition boards

一等奖

迈克尔·迪克森（澳大利亚，布里斯班）

这个设计是个大胆的尝试，整体规划由若干多层结构组成，与大多数高密度城区外的新学校都不同。但是评审考虑了这个设计的积极方面——它将各种各样的元素组成了一个整体。
从哲学方面思考，设计师所提出的一些理念更适合一所标准城际公立学校，而不是这样一所专注于传统高等教育的学术院校。灵活性是这个设计的主要方面，双层表皮结构提供了最大限度的变化度，不仅可以重新布置家具，还能在结构上进行重心调整。——斯坦利·科利尔

本页和对页
竞赛展示板

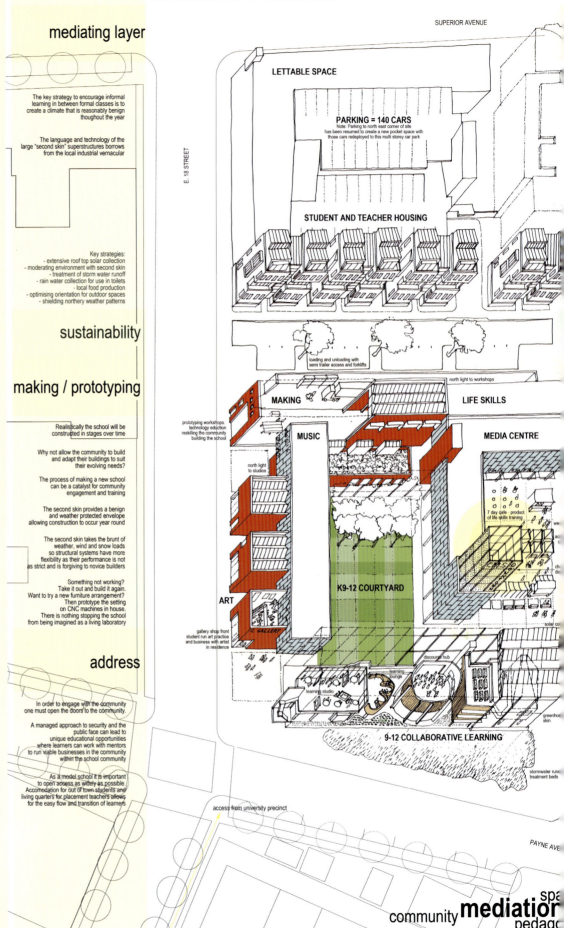

SITUATIONAL FRAMEWORK

mediating layer

The key strategy to encourage informal learning in between formal classes is to create a climate that is reasonably benign thoughout the year

The language and technology of the large "second skin" superstructures borrows from the local industrial vernacular

Key strategies:
- extensive roof top solar collection
- moderating environment with second skin
- treatment of storm water runoff
- rain water collection for use in toilets
- local food production
- optimising orientation for outdoor spaces
- shielding northery weather patterns

sustainability

making / prototyping

Realistically the school will be constructed in stages over time

Why not allow the community to build and adapt their buildings to suit their evolving needs?

The process of making a new school can be a catalyst for community engagement and training

The second skin provides a benign and weather protected envelope allowing construction to occur year round

The second skin takes the brunt of weather, wind and snow loads so structural systems have more flexibility as their performance is not as strict and is forgiving to novice builders

Something not working? Take it out and build it again. Want to try a new furniture arrangement? Then prototype the setting on CNC machines in house. There is nothing stopping the school from being imagined as a living laboratory

address

In order to engage with the community one must open the doors to the community.

A managed approach to security and the public face can lead to unique educational opportunities where learners can work with mentors to run viable businesses in the community within the school community

As a model school it is important to open access as widely as possible. Accomodation for out of town students and living quarters for placement teachers allows for the easy flow and transition of learners

community mediation space pedagogy

PEDAGOGICAL FRAMEWORK

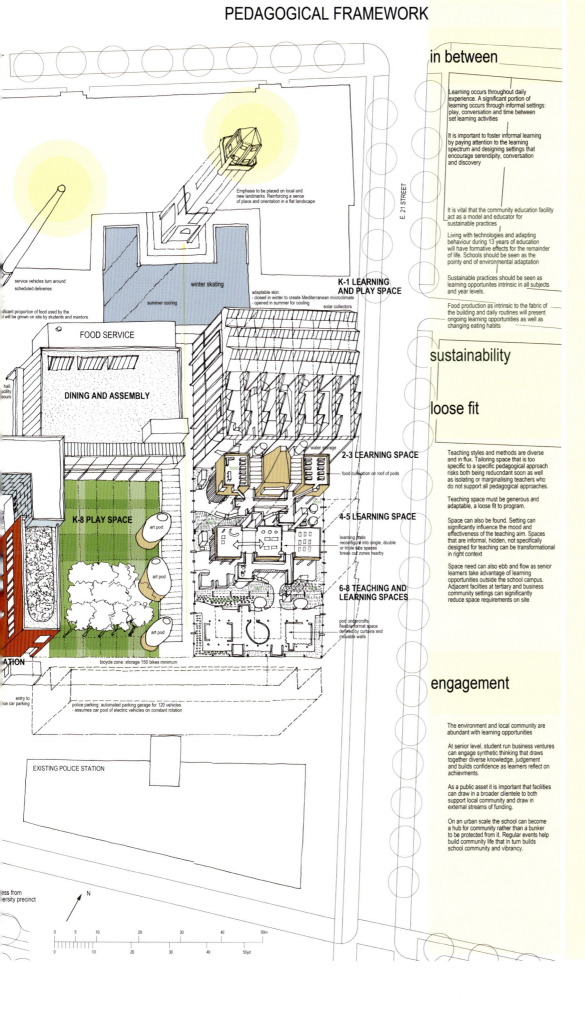

in between

Learning occurs throughout daily experience. A significant portion of learning occurs through informal settings: play, conversation and time between set learning activities

It is important to foster informal learning by paying attention to the learning spectrum and designing settings that encourage serendipity, conversation and discovery

It is vital that the community education facility act as a model and educator for sustainable practices

Living with technologies and adapting behaviour during 13 years of education will have formative effects for the remainder of life. Schools should be seen as the pointy end of environmental adaptation

Sustainable practices should be seen as learning opportunites intrinsic in all subjects and year levels.

Food production as intrinsic to the fabric of the building and daily routines will present ongoing learning opportunities as well as changing eating habits

sustainability

loose fit

Teaching styles and methods are diverse and in flux. Tailoring space that is too specific to a specific pedagogical approach risks both being reducndant soon as well as isolating or marginalising teachers who do not support all pedagogical approaches.

Teaching space must be generous and adaptable, a loose fit to program.

Space can also be found. Setting can significantly influence the mood and effectiveness of the teaching aim. Spaces that are informal, hidden, not specifically designed for teaching can be transformational in right context

Space need can also ebb and flow as senior learners take advantage of learning opportunities outside the school campus. Adjacent facilities at tertiary and business community settings can significantly reduce space requirements on site

engagement

The environment and local community are abundant with learning opportunities

At senior level, student run business ventures can engage synthetic thinking that draws together diverse knowledge, judgement and builds confidence as learners reflect on achievments.

As a public asset it is important that facilities can draw in a broader clientele to both support local community and draw in external streams of funding.

On an urban scale the school can become a hub for community rather than a bunker to be protected from it. Regular events help build community life that in turn builds school community and vibrancy.

Home School Home

Home School Home is an urban strategy to use Cleveland's failing housing stock to re-imagine the public school.

Cleveland's foreclosure crisis has a direct impact on the success of it's schools.

The combination of a deep recession and the foreclosure crisis has left the Cleveland Metropolitan School System struggling financially. Home School Home uses the program and resources of the school to fight this cycle. By siting the school directly in neighborhoods devastated by the foreclosure crisis, Home School Home stabilizes the neighborhoods' housing markets, in turn stabilizing the long term funding of the school system.

Since 2009 foreclosures have resulted in a total loss of $576.6 million in surrounding property values.

1 in 13 Cleveland homes are considered long term vacant properties.

Managing Cleveland's vacant foreclosures costs the city $2.4 million per year in direct expenses.

Annually, these vacant properties result in an loss of $11.2 million in property tax revenues.

Home School Home is an urban strategy to foster the cultural and educational life of Cleveland.

The schools create a new cultural promenade along Superior Avenue that can be experienced by a broader range of people and brings the unique elements of the school to the community where they belong. In Home School Home, the school is dispersed throughout the city in a much finer grain than usual. The typical large school is broken into 12 small schools to maximize the impact in the neighborhoods of Cleveland.

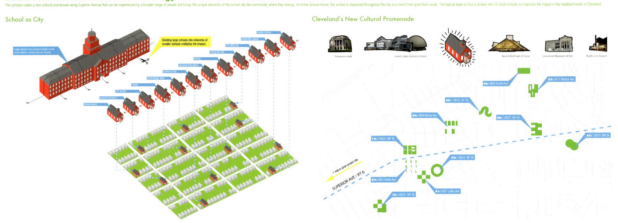

Home School Home is an amalgamation of house and classroom.

Home School Home is able to easily adapt to existing residential properties because of its innovative approach to the classroom that proposes that the total student body of 780 is broken down into 12 smaller groups of 65 students each with a representation of the entire range of K-12 in each school. Throughout a child's K-12 education, he not only experiences 12 different school buildings, he also experiences 12 different areas of the city.

Home School Catalog

01
Plus Lab
536 W 64ᵗʰ Street

02
Art Loop
1363 W 65ᵗʰ Street

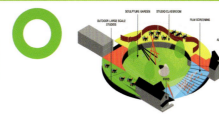

Second Place

Michael Robitz, Sean Franklin, Alexandra VanOrsdale
New York, NY

As a socially imaginative proposal, this entry was as much about preserving the community fabric, in both a physically and holistic sense, as it was about improving school performance. Instead of sticking to the designated competition site, the authors used an entire neighborhood for their plan, targeting abandoned dwellings as learning centers.

Of course, there would be a lot of logistical problems with such an arrangement; and attracting a certain percentage of suburbanites to such a school would certainly assume a lot of broadmindedness on the part of the latter-in a city where segregated housing has been the norm for decades. Still, this entry suggests how a decentralized arrangement of of a physical environment can be woven into a viable learning process. -SC

03
Dewey Decimal Garden
3901 Dibble Ave

04
Home School
8502 Bonna Ave

05
Planetarium Bubble
1337 E 89ᵗʰ Street

06
Practice Field
8002 Crumb Ave

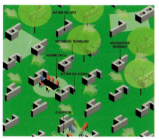
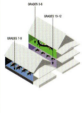

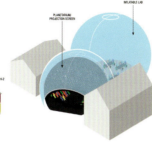
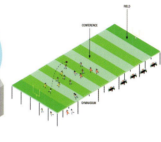

THIS PAGE AND OPPOSITE
Competition boards

07
Home Theater
8117 Medina Ave

08
Curated Classroom
1242 E 82ⁿᵈ Street

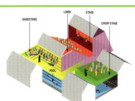

09
WIFI Walkway
1163 E 72ⁿᵈ Street

二等奖

迈克尔·罗比兹；西恩·富兰克林；亚历山大·范尔斯达尔（纽约）

作为一件富有创造力的作品，这个设计在自身和整体方面都保护了社区结构，旨在促进教育发展。设计没有局限于指定的竞赛场地，利用整个街区作为它们的设计场地，将废弃的住宅作为学习中心。

当然，这里有许多统筹问题，例如布局；而且吸引大量城郊居民到学校就学也会产生许多问题。而且这个作品展现了一个分散的物理布局是如何被编织进可行的学习进程之中的。——斯坦利·科利尔

本页和对页
竞赛展示板

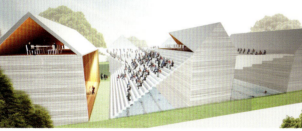
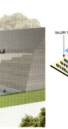
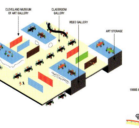

10
Vocational Garages
1268 E 49ᵗʰ Street

11
Tax-base Towers
18th & Payne Ave

12
Edible School Yard
1268 E 59ᵗʰ St

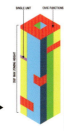

New School Vision
新学校设计

Third Place (2)

Drozdov & Partners
Kharkov, Ukraine

ABOVE
View to plaza with downtown
Cleveland in the background
RIGHT, MIDDLE
Floor plan
RIGHT, BELOW
Section
OPPOSITE PAGE, ABOVE
Interior perspectives
OPPOSITE PAGE, BELOW
Competition board with aer-
ial view of site

三等奖（2）

德罗兹多夫事务所（乌克兰，哈尔科夫）
上图
广场和市中心，背景是克利夫兰城
右中
平面图
右下
剖面图
对页，上图
室内景象
对页，下图
竞赛展示板，场地鸟瞰图

the school garden

Third Place (2)

Drozdov & Partners Ltd.
Oleg Drozdov
Anna Kosharnaya
Pavel Zabotin
Andrian Sokolovksy
Kharkov, Ukraine

If you were able to build a total package in one fell swoop, this might be the most appealing choice of all the entries. Building it in phases would hardly seem to be an option; but the emphasis on nature and sustainability makes it a viable learning experience in itself. Light wells are used to bring sunlight into the interior of the structure, and the more traditional drop-off area at the entrance gives little hint of what is to transpire once one is inside. In a very intensive, and sometimes chaotic urban setting, this provides an element of tranquility, where contemplation cannot be entirely ruled out. -SC

三等奖（2）

德罗兹多夫事务所（乌克兰，哈尔科夫）

如果你能够在刹那之间建造完整的方案，这也许就是最吸引人的设计。项目的分期建设几乎不可行，但是其在自然和可持续性上的着重点让项目值得借鉴。天井的设计为室内带来了阳光，而更加传统的入口下客区预示了室内的场景。在密集而喧嚣的城市环境中，这个项目提供了一个宁静的场所，让人可以静下心来思考。——斯坦利·科利尔

processes

ecology

The project suggests a borderless, human-friendly and clean environment. One-storey volume reduces obstacles, both physical and visual, to minimum and allows the creation of a green plateau. The plateau goes down from the main volume to the courtyard and makes up a hill which serves as a shelter for parking. Climbing up the sloping green roof or the hill, one can enjoy the panoramic view of the city. The inner space of the school forms one continuous whole with no rigidly segregated areas. Educational zones flow one into another. Sliding screens can be moved, and open multimedia libraries and laboratories make up a vast open space which can accommodate various events. Classrooms are located along the perimeter. Fully glazed facade overlooks the street and creates a visual connection between the school and the city.

elevation B'-A"
scale 1:300

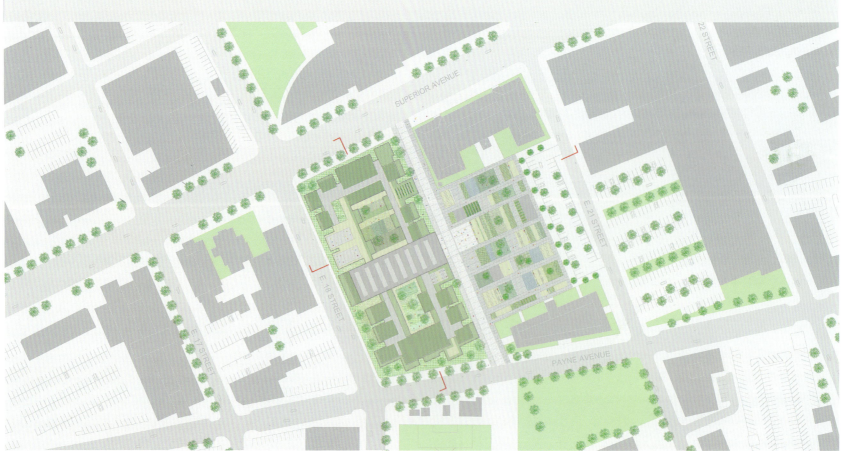

NLIGHTENMENT
1 CLEVELAND DESIGN COMPETITION

Third Place (2)

Feld Architecture
Vincent Feld
Paris, France

This scheme makes generous use of the space, placing the structures(s) and activities in such a way that one is hardly aware of thenearby high-density nature of the neighboring urban setting. The architectural expression is very much of a contemporary, minimalist nature, punctuated by surprises such as a boardwalk, which bisects the campus. In may respects, this school would function quite well in a number of urban environments. -SC

LEFT, ABOVE
View to entrance with downtown Cleveland in background
LEFT, BELOW
Floor plans and organizational scheme
OPPOSITE PAGE, ABOVE
Recreational area
OPPOSITE, BELOW
Site plan

三等奖（2）

菲尔德建筑事务所；文森特·菲尔德（法国，巴黎）

这个设计大量运用了空间，结构和活动的设置几乎让人们忘记了自己置身于一个高密度城市街区。建筑表达十分现代、自然，其间点缀着木板路（它将校园一分为二）等惊喜元素。在许多方面，这所学校在城市环境中都将运行良好。

左上
入口，背景是克利夫兰市中心
左下
平面图和组织结构规划
对页，上图
休闲区
对页，下图
总规划图

FIRST FLOOR

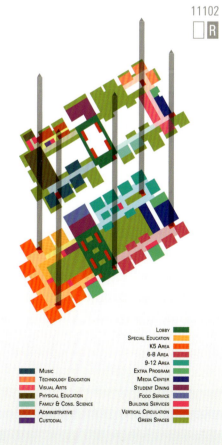

SECOND FLOOR

11102

Music	Lobby
Technology Education	Special Education
Visual Arts	K5 Area
Physical Education	6-8 Area
Family & Cons. Science	9-12 Area
Administrative	Extra Program
Custodial	Media Center
	Student Dining
	Food Service
	Building Services
	Vertical Circulation
	Green Spaces

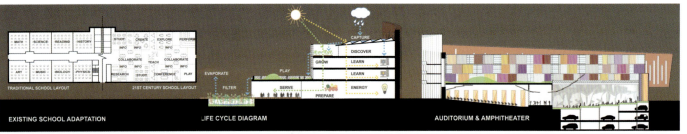

EXISTING SCHOOL ADAPTATION   LIFE CYCLE DIAGRAM   AUDITORIUM & AMPHITHEATER

荣誉奖

温德尔建筑事务所（纽约州，阿默斯特）

本页
竞赛展板中的鸟瞰图和地面展示图

"I HEAR AND I FORGET. I SEE AND I REMEMBER. I DO AND I UNDERSTAND." -CONFUCIUS

Honorable Mention

Wendel Architecture
Amherst, New York
Team
Michael Conroe
Leanne Stepien
Giona Paolercio
Stephanie Vito

THIS PAGE
Aerial and grade level views
from presentation boards

"LOGIC WILL GET YOU FROM A TO B. IMAGINATION WILL TAKE YOU EVERYWHERE." -EINSTEIN

Honorable Mention

KGD Architecture
Rosslyn, Virginia

荣誉奖

KGD建筑事务所（弗吉尼亚州，罗斯林）

上图
学校场地中的行人视角
左图
室内景象

ABOVE
Pedestrian perspective within the school site
LEFT
Interior view

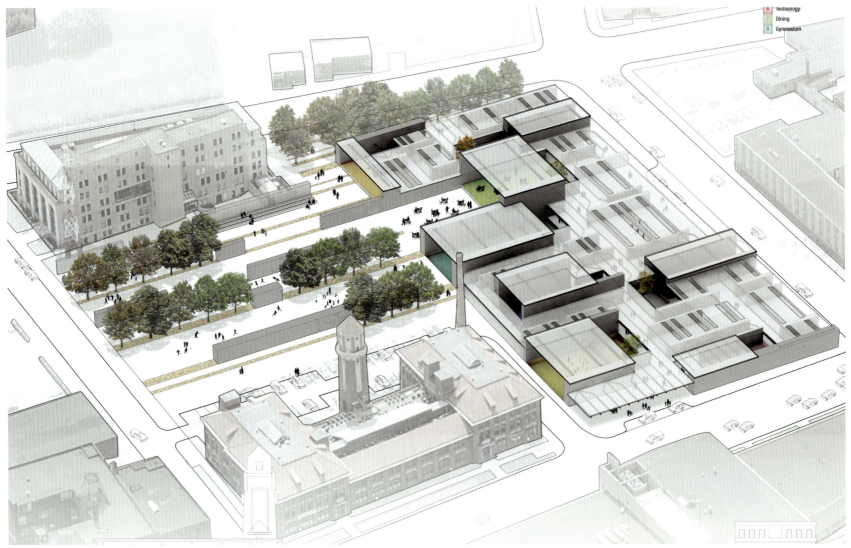

Honorable Mention (opposite page)

Lateral Office

Toronto, Canada
Mason White
Lola Sheppard
Nikole Bouchard
Zoe Renaud-Drouin
Paul Christian
ionn Byrne

OPPOSITE, ABOVE
Pre-school play area
OPPOSITE, BELOW
Aerial view of project

荣誉奖（对页）

横向工作室（加拿大，多伦多）

对页，上图
学前游戏区
对页，下图
项目鸟瞰图

Honorable Mention (above)

Jedidiah Lau

Hong Kong

ABOVE
Large multi-purpose room housing various
activities

荣誉奖

耶底底亚·刘（中国，香港）

上图
大型多功能活动室

Live/Workplace as Tourist Destination:
Fez, Morocco Infill & Planning Competition

5

The U.S. government funding design competitions abroad? Especially when it has almost been absent in supporting such programs at home? It was not too long ago that the U.S. Congress passed a law stating that no federal funds could be used to fund U. S. international expo pavilions, let alone competitions to determine their design. So those who are wondering that our federal government is spending tax dollars on foreign soil to promote good design should know that its funding for for the Place Lalla Yeddouna redevelopment competition in Fez, Morocco was mainly the result of an economic redevelopment grant from the U.S. Government, and that the competition was only a peripheral add-on.

By its very nature, this competition was not the type of project to attract star architects. While the competition concentrated on upgrading a dilapidated neighborhood where a number of structures are to be replaced, this project was more about giving a boost to local economic development than it was about visuals. Here the existing housing stock, built primarily of masonry, went a long way in determining the parameters of the infill program. Moreover, the site has been a UNESCO world heritage site since 1981. So there was little emphasis on the super modern, especially in view of the constraints resulting from the very nature of the site. As a result, the usual cast of

star architects was nowhere to be seen. Rodolfo Machado, a juror and frequent participant in foreign competitions, stated that "the usual luminaries were missing, and I didn't recognize any of the participants."

美国政府竞赛为海外设计竞赛提供资金？它还几乎没有为国内此类竞赛提供资金呢。不久之前，美国国会通过了法案，不允许政府使用联邦基金来资助海外展馆，更不用提设计竞赛了。然而，拉拉雅德那地区重建竞赛的资金主要来自于美国政府为摩洛哥提供的重建资金，竞赛只是附加条件而已。
这次竞赛的性质注定了它不会吸引知名建筑师。竞赛聚焦于提升破败的街区，拆除一些结构，借此推动本地经济发展。原有的住宅主要由石材建造，对决定项目参数相当有帮助。此外，竞赛场地是世界文化遗产。因此，场地上的建筑不能过于现代，从而也就不能吸引明星建筑师。评审鲁道夫·马查多称："常见的杰出人物都缺席了，我不认识任何一位参赛者。"

一等奖（55,000美元）

摩塞西安事务所（英国，伦敦）和亚瑟尔·卡里尔工作室（摩洛哥，卡萨布兰卡）

对页
模型鸟瞰图

下图
剖面图和立面图

1st Prize (US$55,000)

Mossessian & partners
London, UK
with Yassir Khalil Studio
Casablanca, Morocco
Team
Anthony Mopty
Mourad Bellaanaya
Abdelkarim Tounli
Saana El Kahlaoui
Nadia Azia
Freelance collaborators
Maurel Fabrice, Architect
Youssef Lahrichi, Architect

OPPOSITE PAGE
Aerial View of model

BELOW
Section and elevation

The jury was made up of seven architecture experts and six technical experts. The architecture-related experts were:

- Matthias Sauerbruch, Berlin, Germany (Chair)
- Marc Angélil, Zurich, Switzerland
- Meisa Batayneh Maani, Morocco
- Mohamed Habib Begdouri Achkari, Morocco
- Dr. Stefano Bianca, Geneva, Switzerland
- David Chipperfield, London, U.K.*
- Rodolfo Machado, Boston, USA

With the exception of Richard Gaynor of the Millennium Challenge Corporation (Washington, DC), the expert jurors were mostly local: Youssef El Mrabet, Mohamed Hassani Ameziane (Architect), Mohammed Msellek, Fouad Serrhini and Abdelkader Nadri (Architect).

•David Chipperfield was only present for the first stage of the proceedings.

The Judging Process

Conducted in two stages, the discussion focussed on four entries in the final phase of adjudication. According to the jury report, "expert jurors and Institutional representatives" remarked on the on the "variety and inventiveness of the submitted proposals and acknowledged that the competition had achieved its intended goals." Together with the other jurors they attested to the general depth of investigation and the remarkable quality of the presentations, particularly in light of the relatively short time between the first and second phase of the competition.

The key issues which informed the final decision were the successful integration of a new intervention into the organic fabric of the city, the exploration of the site as a location for contemporary activity, the buildability of the proposals and the apparent professionalism of the team. After much discussion, the jury agreed on the following ranking:

Summary

Although some may find this competition, lacking a major building, short of a real design challenge, there are some lessons to be learned here. Although architects were not asked to design a mega-structure as part and parcel of this program, the plethora of solutions for urban infill should provide all urban planners with food for thought. Although very site specific, the research and thought given to this infill process might well be emulated by planners in other parts of the world. Based on the results of Place Lalla Yeddouna redevelopment, the road to such solutions can reside in carefully planned and executed design competitions.

评委会由七名建筑专家和六名技术专家组成。建筑专家为：

马提亚·绍尔布鲁赫（德国，柏林）（评委会主席）

马克·安格里尔（瑞士，苏黎世）

梅萨·巴泰尼尔·马阿尼（摩洛哥）

穆罕默德·哈比卜·柏格杜里·阿卡里（摩洛哥）

斯特凡诺·比安卡（瑞士，日内瓦）

大卫·科波菲尔（英国，伦敦）

鲁道夫·马查多（美国，波士顿）

出于世纪挑战集团（华盛顿）理查德·加诺尔的期望。专家评委基本都来自当地，他们分别为：

约瑟夫·埃尔·马拉别特、穆罕默德·哈萨尼·阿美齐亚（建筑师）、穆罕默德·马瑟里克·福阿德·赛尔里尼和阿布德卡德·纳德里（建筑师）。大卫·科波菲尔只参与了第一阶段的评审过程。

评审流程

竞赛分为两个阶段，最后决定出了四件入围作品。根据评审的汇报："专家评审和学会代表们"对"参赛作品的多样性和创造性"印象深刻，并且"认为竞赛实现了既定目标"。他们与其他评审一起，在短时间内对作品质量进行了评估。最终决定竞赛结果的重要因素在于项目是否能够成功地融入有机城市网格、项目对场地的探索程度、项目的可行性以及设计团队的专业精神。

总结

尽管一些人认为本次竞赛缺乏主要建筑和真正的设计挑战，这里还有许多经验教训值得学习。尽管竞赛没有要求建筑师打造一座巨大的建筑结构，多种多样的城市填充规划将为城市规划者提供精神食粮。尽管竞赛场地具有局限性，填充过程中所进行的调研和思考还是可以被运用到世界各地。以拉拉雅德那地区重建为基础，这类解决方案能够被执行在精心规划的设计竞赛之中。

2 PLACE LALLA YEDDOUNA

010067

SITE PLAN | SCALE 1:200 @ A0

e Features

TRANSVERSE SECTION A | SCALE 1:150 @ A0

1st Prize (US$55,000)

Mossessian & partners
London, UK
with Yassir Khalil Studio Casablanca, Morocco

Jury Comments

At first glance, the winning entry by Mossessian & Partners might place it in the category of the "traditional footprint." Upon closer examination, their scheme showed a sophisticated understanding of the program, using a network of paths and passages to link the main site with both the city and the river, "integrating the riverbanks into the traditional street network of the Medina." Aside from the main access streets from the rest of the city, the internal passages linking small courtyards were not in a continuous linear configuration, thus creating a more interesting flow, and an intimate spatial setting for local residents. The jury also felt that "the simple cuts in the buildings and color-line gateways were a powerful instrument to introduce contemporary spaces into the scheme.

Additionally, the jury agreed on "the supreme feasibility of this complex urban form within the constraints of local building methods and materials." The materials submitted attested to the high level of professionalism of the team, and the restoration techniques and attention to construction details gave every indication that the project could be built well within time and budget.

一等奖（55,000美元）

摩塞西安事务所（英国，伦敦）和亚瑟尔·卡里尔工作室（摩洛哥，卡萨布兰卡）

评审评语

乍一看，这件优胜设计可能会被划分到"传统规划"之中。经过仔细审视，项目将逐步展示出设计师对项目的完全了解。他们利用一个道路网络将主场地与城市和河流连接起来，"将河堤整合成麦地那式的传统街道网络"。除了进入城市的主要街道之外，连接各个庭院的内部走道都没有采用连续的直线结构，从而形成了更有趣的流动和私密空间布局。评审还认为"简单的建筑切割和彩线通道为项目带来了强有力的现代空间"。

此外，评审们一致赞同"这个城市规划在限制条件下的可行性"。材料的运用展现了设计团队的专业水平，而修复技术和对建筑细部的注意让项目能够在预定时间和成本内建成。

REHABILITATION STRATEGY

TRADITION AND INNOVATION:

These precious buildings represent invaluable examples of Moroccan building and urban typologies; thus the process of preservation, reconstruction and insertion requires an understanding of the essential elements that make up this typology. Insertion of new construction is in the form of crafted furniture pieces that benefit the programme and also partition the spaces for their appropriated uses. An effort is made to create minimum physical joinery between the preserved and inserted elements. The same attitude follows throughout the entire site - where the new respectfully meets the old.

A unified strategy is applied to all preserved buildings but with regards to their typologies: the "Dar" and the "Fondouk." As in the case of building I-02, "Dar" type consists of large rooms that are laid out around a central courtyard. The "Fondouk" type (example Building I-01) is a series of small shops which surround a smaller courtyard. Additions and transformations that were done during recent history would be removed in order to revert back to the original layout of building types. Their structure will be reinforced and partitions, where possible, will be removed. Existing circulation elements will be repaired and if possible be enlarged to allow the vertical transportation of tools and artworks. In the case of "Dar" type, furniture units will reinforce the boundaries of the rooms, and in "Fondouk" these units will be arranged to form cellular spaces.

The furniture elements will provide the necessary work stations for the artisans. The elements will contain foldable desks, lockers for tools and shelves to store and display the works. An uninterrupted circulation zone between artisan's space and the courtyard will be maintained so that the visitors will have the chance to see the process of making the products which they are about to purchase.

The open space configuration will help the artisans to interact with each other as well, reinforcing the sense of community. Furthermore, the overall production will benefit through the share of knowledge and craftsmanship. We envision these spaces as a new base for a united community of artisans that work, socialize and share their knowledge with new generations to come.

LINKED COURTS:

Adjacent courts can be linked where possible and flexible. Thresholds will be introduced with respect to the layout and materiality of courtyards. This will allow visitors and artisans to walk through the old quarter without the need of exiting back to street level.

DAR
Existing Condition

FONDOUK
Existing Condition

New Strategy

New Strategy

DAR TYP: BUILDING I-02

FONDOUK TYP: BUILDING I-01

RIVER LEVEL PLAN | SCALE 1:150 @ A0

SECTION THROUGH THE SOUQ | SCALE 1:150 @ A0

1st Prize

Mossessian & partners
London, UK
with Yassir Khalil Studio
Casablanca, Morocco

LEFT
Competition board 5
OPPOSITE PAGE
Competition board 7

一等奖（55,000美元）

摩塞西安事务所（英国，伦敦）
和亚瑟尔·卡里尔工作室（摩洛哥，卡萨布兰卡）

左图
5号竞赛展示板
对页
7号竞赛展示板

UPPER LEVEL PLAN | SCALE 1:150 @ A0

0m   5m   10m

TRANSVERSE SECTION B | SCALE 1:150 @ A0

Stone mason   Weaving shop   Silversmith's   Carpentry

PLACE LALLA YEDDOUNA WEST ELEVATION | SCALE 1:150 @ A0

0m   5m   10m

二等奖

费雷蒂·马尔赛罗尼（意大利，罗马）

2nd prize

Ferretti-Marcelloni
Rome, Italy

二等奖设计以其"良好的空间整合"深受评审好评。它巧妙地在现有场地中融入了现代语言，同时也保持了本地风格和材料。设计师运用河岸的"硬性"垂直边缘，不仅为项目提供了高度密集型，也让人们在冬天能够体验外面河岸上的风景。一系列的可持续特征包括：利用喷泉来减少夏天的热源和光线结构，提升室内的活跃度。在项目的建筑表达上，评审一针见血地指出："它拥有始终如一的造型语言，而又不单调。项目不仅提升了现有场地，它的北美式建筑语言还促进了现代建筑表达。"——斯坦利·科利尔

本页
模型鸟瞰图和效果图
对页
总规划图和行人视角

The second-place Ferretti-Marcelloni entry received high points from the jury for its "good volumetric integration into the historic urban fabric, the clever use of existing site opportunities and the implementation of a contemporary language that is based on local typologies and materials. Using a "hard" vertical edge at the river's edge not only provides the project with a sense of high density; the way in which it is penetrated from the river might suggest a sense of discovery once one is inside those walls and a different level of sophistication in urbanity from the outright winner. There are a number of sustainability features, including the use of fountains to lower the heat factor in the summer and light shafts in the structures, enhancing livability in the interior. When speaking of the project's architectural expression, the jury hit the nail on the head by describing it as having "a consistent formal language without being monotonous. Here it was not just about upgrading the existing; the architectural language here is North African in massing, but pushing toward the modern in architectural expression. -SC

THIS PAGE
Birdseye view of model and perspective
OPPOSITE PAGE
Site plan and pedestrian view

The relationship between the urban tissue and the still uncovered stretch of the river, is characterised by a urban compact front.

Inside the project area is proposed a tighter alternation of continuos elements an passages, with the purpose of define a new per ceptive relationship between the Medina and the river. Indeed from Place Lalla Yeddouna, will be possible to see the Oued's course.

Moreover Place Lalla Yeddouna represents an urban rest, in between the way from the Jamaa Andaluse Mosque to the Qarawiyine Mosque; hereby the choise not to treat in a continuos way the front of the path, to highlight the relationship between the two centers.

OUED FRONTS

Inside a complex sistem of nodes and landmarks – El Seffarine Square, R'cif Mosque, Jamaa Andalous Mosque, Quarawiyine Mosque and many others– is included Place Lalla Yeddouna as a center of the Artisan Circuit and place accessible to the tourists. The idea of the project is to organize the different functions respecting th e existent morfology with the purpose of encourage a flexible, mixed use of this part of the city.

LANDMARKS

Place Lalla Yeddouna has always been excluded from the main Tourist Circuits; the intention to requalificate the Oued is realized with the introduction of a new vertical axis of penetration in the urban tissue, able to connect R'cif to Place Lalla Yeddouna-currently two of the voids more in proximity of the main car accesses.
The project suggest The Oued as a connection axis from wich you can reach the pedestrian paths of the Medina and able to define a new linear space of re-connection between the two centers.

LINEAR CONNECTION

Area Plan 1:1000

The contemporary image of the Medina : a nocturnal view - scale 1:200

The openings and the passages in the new block towards the Oued and the new pathways mutually create a functional and visual dialogue

The contemporary image of the Medina : a diurnal view - scale 1:200

2nd prize

Ferretti-Marcelloni

Rome, Italy

LEFT
Elevations and Sections
OPPOSITE PAGE, ABOVE
View of bridge connector and creek bed
OPPOSITE, BELOW LEFT
View from above of rooftops
OPPOSITE, BELOW RIGHT
Detail of grade plan

二等奖

费雷蒂·马尔赛罗尼（意大利，罗马）

左图
立面图和剖面图
对页，上图
桥梁和河床
对页，左下
屋顶鸟瞰图
对页，右下
平面规划细部

Section DD' - scale 1:200

Section CC' - scale 1:200

Section AA' - scale 1:100

Section BB' - scale 1:100

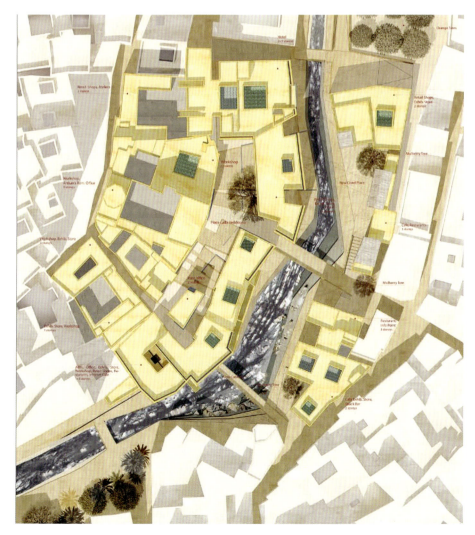

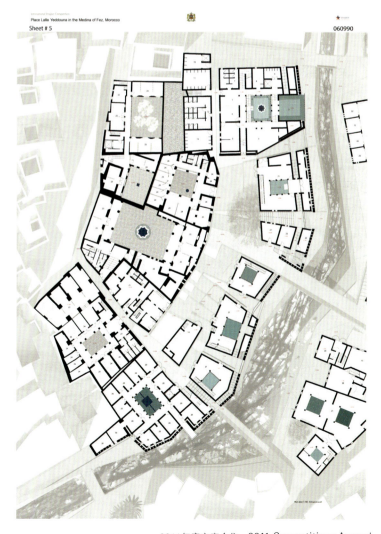

3rd prize:

Moxon Architects
London, UK

三等奖

默克森建筑事务所（英国，伦敦）

默克森建筑事务所对现代文化中的历史保护理解深刻。他们将"巧妙"作为主要原则，连接了新与旧，聚焦于调节的工艺。"原有建筑被尽量保留，新建筑与场地相融合——这让整个规划进一步修复了麦地那的原始城市网格。"与优胜设计相同，这一设计也十分注重公共、半公共和半私密区域的交通组织。另一方面，评审发现"项目细部的一些设计脱离了环境，过于复杂、'洁净'或'冷静'"。总体来说，设计打造了"一个特别有趣的解决方案，呈现了一个价值极高的城市规划策略。"——评审评语

本页
带有效果图的竞赛展示板
对页
模型鸟瞰图和细部图

Moxon Architects was praised for its examination of the role of historic preservation within contemporary culture. They emphasized 'subtleness' as a principle posture in relating the new and the old, with a focus on the craft of intervention. "Existing buildings are retained where possible and new building fit preexisting footprints — in this way the scheme is above all else one of conscientious repair to the precious fabric of the medina." As with the winner, the care with which the organization of the circulation pattern into public, semi-public, and semiprivate was in the forefront.

On the other hand, the jury found that "certain design propositions at the detail level appear to be out of context and partially overdone in terms of the composition of elements and their 'cleanness' or 'coolness' in style. Overall, the proposal offered "an extremely interesting solution to the competition brief and presents a highly valuable strategy for addressing the transformation of historic urban fabric.
-Jury Comments

THIS PAGE
Competition board with renderings
OPPOSITE PAGE
Aerial views of model with detail

3rd prize:

Moxon Architects
London, UK

THIS PAGE
Competition board showing
sections and elevations
OPPOSITE PAGE
Illustrations indicating use of
materials

三等奖

默克森建筑事务所（英国，伦敦）

本页
竞赛展示板，显示了立面图和剖面图
对页
材料运用示意图

340088

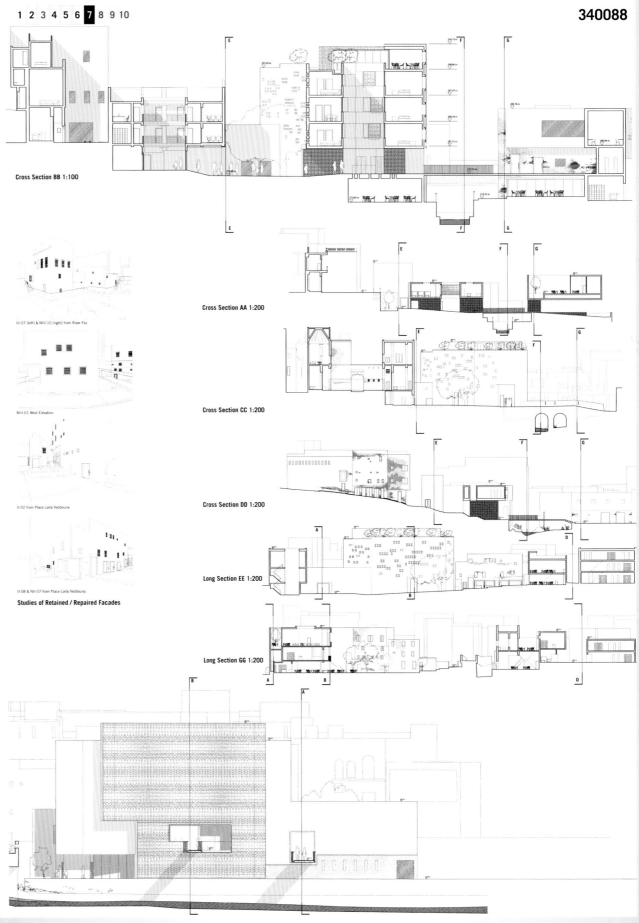

Cross Section BB 1:100

III-07 (left) & NIII-10 (right) from River Fez

NIII-01 West Elevation

II-02 from Place Laïla Yeddouna

II-08 & NII-07 from Place Laïla Yeddouna

Studies of Retained / Repaired Facades

Cross Section AA 1:200

Cross Section CC 1:200

Cross Section DD 1:200

Long Section EE 1:200

Long Section GG 1:200

Bridge link (through to North end of site) [NH-11]

Proposed bridges across River Fez

New public landscape to east bank of the River Fez, looking towards Bim Lamdoum Bridge

East abutment of Bim Lamdoum across to west bank and Place Lalla Yeddouna beyond

Bim Lamdoum towards new bridges (north end of site)

Window grille studies [currently exist on the site]

Pressed, polished and cut screen to hotel facade based on Berber craft patterns. Apertures define window areas and solid areas define the location of solar thermal collectors for the site wide provision of hot water.

Facades

Reinforced concrete will be used throughout the site for both new buildings and repairing existing buildings. Reinforced concrete (RC) is the obvious choice in material as it is used intensively throughout Fez, including this site and hence the new structures will be in keeping with the existing. The site constraints also make RC a good choice in material as it will be easy to get into and around the site (compared to say long pieces of fabricated steel) and required levels of technical ability commensurate with those available locally. Other advantages include:

It can be made using local materials, thus providing a cost effective solution and encouraging local development.

RC requires low maintenance, which is important for both cost and access.

It provides thermal insulation against the heat.

RC is durable in the Moroccan climate.

In addition to the general use of massive construction techniques (surface concrete, rendered blockwork and facing brick) the proposal makes limited but prominent use of cut and etched bronze and steel sheet as a facade material defining both solid wall and window opening. In common with the approach to veneers and linings the patterning is based on local precedent, however rather than following the direct arabic antecedent of decorated entrances, the metal elements follow the dual precedent of wrought iron window grilles (visible throughout Fez) and the patterns of traditional Berber weaving and carpet making.

Form

The building forms proposed are developed with a good deal of commonality with the pre-existing layout on the site. From this starting point their development in massing forms has been undertaken to consider the total effect on the internal environment and adjacent external environment along with the energy in use implications of these forms. There are inherent interrelationships with optimum orientation for building integrated renewables, daylighting and shading which have been considered to this stage in the development of the design. Specifically the proposal represents the development of the following ideas:

Optimise room orientation to best suit the function of rooms in order to improve the occupant's environment and reduce energy use where possible.

Exploit shading from adjacent developments and self-shading to reduce direct solar gains. The scheme, while developed from the existing urban layout, makes extensive use of projecting volumes and self shading in the building forms.

External spaces such as balconies, terraces or accessible roof areas are shaded. The shading presents an opportunity to incorporate solar thermal or photovoltaic panels. This opportunity has been taken where possible, with the South East elevation of the hotel forming a large scale solar thermal collector and the roofscape of many of the new buildings being taken up with photovoltaic panels.

The form of the hotel in particular, with the tall central courtyard has been developed to harness the prevailing winds for cross ventilation.

Retained and refurbished buildings and facades

Refurbished & new Courtyards, Linings and Decorations

Layering

The Makina and Fondouks centered around Place Lalla Yeddouna constitute an extraordinary urban scene of artisan industry and commerce. The connection between the customer and craftsman is of a palpable and direct intensity. For the area to retain this vital quality it is important that the mix of manufacture and retail continues to pervade the whole site - including by having an influence on the spatial arrangement of the new hotel and restaurants.

In situating a hotel on Place Lalla Yeddouna care needs to be taken so as not to isolate it or its patrons from the life of the artisan quarter. To this end the proposal incorporates supporting workshops and Artisan exhibition stores in the body of the hotel building, commingling artisans with hotel guests in order that a stay in the hotel is to all intents a stay in the Makina.

The central idea to the planning of the hotel building is that it is at least in part a Fondouk in its own right - sleeping above merchants stores. Indeed the prospect of visiting makers and designers staying at the hotel in order to collaborate with local artisans reinforces this idea of a Fondouk. Guests without a vocational connection to the local industry will nonetheless have the benefit of being embedded in the culture of the place.

The architectural strategy proposed in this scheme is by and large a precise but simple one of repair and replacement. In contrast to this relatively modest approach to the built fabric, the programmatic approach in terms of the economic and social requirements is a bold and emphatic one of maximising the exposure of retail venues to tourist trade, and increasing the opportunities for genuine and meaningful interaction between the local and visiting populations.

Benefits of this interaction include improved access to international trade, better opportunities for enterprise and a greater social connection between people of different backgrounds and nationalities.

The current experience of walking across Bim Lamdoum bridge and clearly hearing, but not seeing, artisan activity in the buildings at its abutments is a uniquely theatrical introduction to the craft work that takes place on the site. The proposal aims to retain the experiential qualities for the visitor by positioning workshops along or adjacent to the principle entrances to Place Lalla Yeddouna. Access to workshops will be controlled but the opportunity for visual and acoustic links to the public areas will be retained and developed.

Existing and proposed Routes, Passages and Bridges

New Construction - both within and separate to retained facades

3rd prize:

Moxon Architects
London, UK

THIS PAGE
Competition board
OPPOSITE PAGE
Site plan

三等奖

默克森建筑事务所（英国，伦敦）

本页
竞赛展示板
对页
总规划图

Central courtyard of proposed building [NII-11]

View from new courtyard, through proposed building [NII-11] and beyond [new bridge & NII-01]

Central courtyard of new Hotel [NII-10]

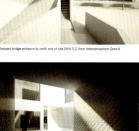

Ramped bridge entrance to north end of site [NII-01] from Interconnection Zone A

Courtyard [NII-11]

Street Pattern

The external sequencing of passages, courts and alleys in the proposal has Arabic and indeed particularly Moroccan antecedents. The circulation pattern is derived from the existing layout so that new bridges and passages have a legible and uncomplicated relationship with the pre existing geometry of the site. The potential richness of these new passages is experienced in the spatial sequencing that they allow — for example from Place Lalla Yeddouna through a short covered alley to a tall narrow courtyard open to the sky, taking a shifted trajectory to another alleyway and then onto a bridge one storey above ground level before descending a set of steps concealed within a narrow slot in the wall of a new building. The key intention in this sequencing has been to cultivate the 'shifted trajectory'. This is in part a consequence of following the pre existing geometry on the site but also importantly a deliberate ploy to introduce the idea of the site being revealed through exploration and movement. By staggering the routes across the site destinations are only revealed through movement; in addition to the pockets of opportunity created by doing so the routes are only understood by walking them – you cannot see from one end to the other. This is akin to the overriding urban form of the Medina of Fez where the place is understood a posteriori through actual use rather than a priori through the comprehension of an overarching organizing structure.

This spatial sequencing also provides for unique opportunities at the intersection between new geometry and existing [retained] fabric. Tightly defined pockets of public space are included across the site at the meeting points between historic and new architecture, and between large scale public domain [Place Lalla Yeddouna] and new discrete niches and circulatory routes. These spaces become the locations of new water fountains, seating and informal congregation. If the archetypal street in the medina of Fez can be characterized by the inclusion of a Mosque, Hamam, Fountain and Bakery the new links in this proposal can each be considered as a modest sequence of mosque, fountain, meeting place and bitter orange tree.

Planting

The Mediterranean climate of Morocco combines the warmth of the desert with the cool breezes of its coastal areas. The position of Fes in the Middle Atlas means that the climate is suitable for a wide range of the country's indigenous plant species – many of which are unique to the Mediterranean and many of which have an obvious appeal for use in the Place Lalla Yeddouna project. In particular the proposal makes use of plant species that have relevance to local craft traditions and confer an aromatic environmental benefit on the public space.

The landscape design is multilayered and in addition to the visual and environmental amenity provided by planting is concerned with yield and scent as organising principles. Examples of species in the scheme include pennyroyal, verbena, thyme, wintwered rosemary and hibiscus, along with Bay, Bitter Orange and Olive trees. In all instances plants are specified that have either medicinal, craft or other commercial value, and are appreciated for their aromatic essential oils as much as their visual amenity value.

Planting is integrated into the architecture. Outdoor seating areas, parapet walls and benches will all be designed as planting opportunities where scent is released when brushing past. In concert with the new city environmental and water cleanliness policies this landscape scheme will contribute to a positively aromatic environment around the Oued Boukhrareb channel.

Public domain also functions as building circulation

Built volumes overlap & define passageways

River channel becomes fragrant & productive garden

Workshops are fitted out with sacrificial linings

Routes become spaces for meetings & markets

Formal landscape to north end of river channel

Public space refurbished on Place Lalla Yeddouna

Semi wild aromatics & orange trees to south of site

Veneers

Where new insertions are made into the retained buildings, or where routes are formed through new buildings, the architectural language is one of veneers and linings. These linings share the principles of hierarchy evident in the internal and entrance spaces of traditional Arabic architecture, where the public domain is often muted in form and expression whilst private and semi private space is more embellished and decorated. So with the new linings a more highly worked material treatment is proposed to the external aspects of the main buildings.

These linings, each with a dominant colour or pattern, become navigational aids for the benefit of guide walkers, visitors and children.

Traditional approaches to the linings of entrances and courtyards have tended to rely on high quality tiled and mosaic patterns, in combination with detailed fretwork and hand carved timber or plaster moulding. This scheme envisages the continued use of traditional hand crafts and also proposes the complimentary use of CNC [Computer Numerically Controlled] 3 axis routers and laser sintering techniques to allow for evolving and non-repetitive patterns based on algorithmic re-interpretations of traditional Moroccan precedents. In all instances it is the craft of making the patterned surface that is important to the success of the linings – whether created entirely by hand or augmented with 'digital craft' techniques.

Bronze Inlays in Polished Plaster [Top Right]

Shallow tray bronze castings, created using a CNC with polymer modified cement before the whole is brushed down and polished to a durable [and manually repairable] finish. This method would result in a smooth but visually complex surface finish of pattern, reflection and colour.

CNC Formed Plaster [Right]

Panels of high density fibre reinforced plaster, directly milled by CNC 3 axis routers, would constitute the highest resolution and most intricate three dimensional patterning in the scheme. These panels would be manufactured off site and would be directly bonded to the supporting concrete or blockwork wall.

Concrete Cast with Reusable Rubber Lined Forework

The rubber formwork for these elements is itself cast using a laser sintered positive mould. The formwork panels would tessellate, allowing for unlimited deployment of the pattern across comparatively large surfaces of exposed concrete. It is anticipated that this method would use the least complex patterns and would have the lowest level of formal fidelity.

Pre-Cast Concrete Panels

The surface finish - and specular reflective qualities - of the concrete will be determined by the stencil applied pattern of rough and smooth on the formwork. The resultant pattern would only be perceptible at oblique angles where the differences in reflectance becomes apparent. This would be the most subtle patterning employed in the scheme.

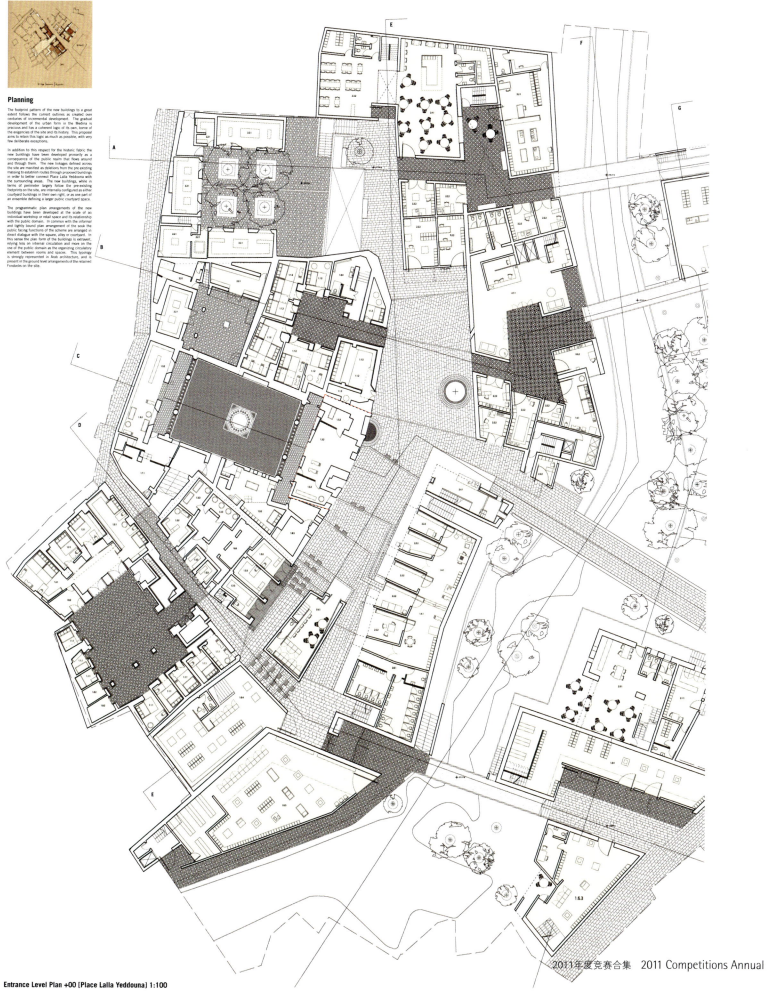

Planning

The footprint pattern of the new buildings to a great extent follows the current outlines as created over centuries of incremental development. The gradual development of the urban form in the Medina is precious and has a coherent logic of its own, borne of the exigencies of the site and its history. This proposal aims to retain this logic as much as possible, with very few deliberate exceptions.

In addition to this respect for the historic fabric the new buildings have been developed primarily as a consequence of the public realm that flows around and through them. The new linkages defined across the site are manifest as deletions from the pre-existing massing to establish routes through proposed buildings in order to better connect Place Lalla Yeddouna with the surrounding areas. The new buildings, while in terms of perimeter largely follow the pre-existing footprints on the site, are internally configured as either courtyard buildings in their own right, or as one part of an ensemble defining a larger public courtyard space.

The programmatic plan arrangements of the new buildings have been developed at the scale of an individual workshop or retail space and its relationship with the public domain. In common with the informal and tightly bound plan arrangement of the souk the public facing functions of the scheme are arranged in direct dialogue with the square, alley or courtyard. In this sense the plan form of the buildings is extravert, relying less on internal circulation and more on the use of the public domain as the organizing circulatory element between rooms and spaces. This typology is strongly represented in Arab architecture, and is present in the ground level arrangements of the retained Fonducks on the site.

Entrance Level Plan +00 [Place Lalla Yeddouna] 1:100

Round 2 Finalist
Bureau E.A.S.T.
Los Angeles, USA and Fez, Morocco

第二轮入围奖

E.A.S.T.工作室（美国，洛杉矶和摩洛哥，菲兹）

左图
场地模型鸟瞰图
左下和对页
总规划图，模型图

LEFT
Aerial view of site model
BELOW LEFT AND OPPOSITE PAGE
Site plan, model perspectives

The jury interpreted the project as a very valuable attempt to establish a plausible symbiosis of the new and the existing. On an urban level the reflection of the basic site topography as the generating factor in the development of a sophisticated internal space is as interesting as the proposed coexistence of two public spaces as different as Place Lalla Yeddouna and the contemporary "soft-edge" river park. The jury also much appreciated the attempt to find principles of social intervention in a very demanding context.

评审认为该项目的价值在于尝试在新旧之间建立了看似合理的共生关系。项目在城市规划层面上反映了场地的基本地形条件，以此为基础开发了纯熟的内部空间，富有公共空间的趣味性，在拉拉雅多纳地区和河岸公园之间形成了对比。评审还十分欣赏设计师在高要求环境中探索社交空间的渴望。

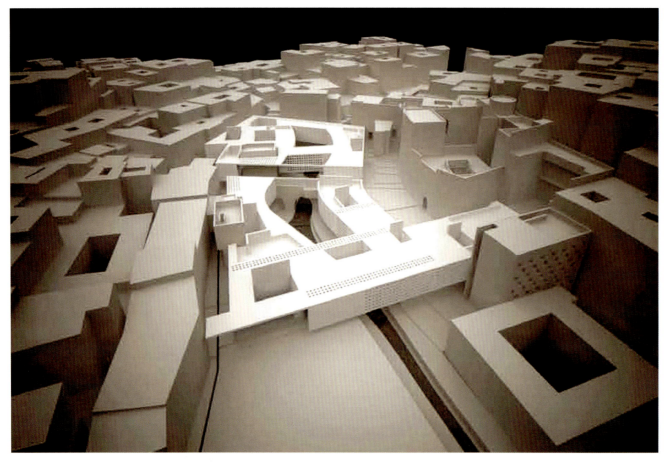

Round 1 Finalist

Kolb Hader Architekten
Vienna, Austria

第一轮入围奖

科博·哈德尔建筑事务所（奥地利，维也纳）

评审欣赏该项目对城市规划的雄心壮志。项目以单一的大型建筑结构为基础，打造了新型的超级建筑。事实上，它是建筑网络的一个碎片，破土而出，进行了具体化。因此，它内部含有诸多庭院和小型街道。

项目的新颖性存在于这种变形之中。项目明晰而精准，为北侧的停车场和拉拉雅多纳提供了大量空间。评审特别欣赏具有公共潜力的屋顶景观。然而，项目的可行性深受质疑，因此评审没有为它颁奖。——评审评语

左上
模型鸟瞰图
左图
场地鸟瞰图

The jury values the ambitious urban strategy of the project. It is based on a single, large architectural intervention, in a way, a new type of mega structure, being no object building. In fact, it is a segment of an architectural tissue raised above the ground and objectified – turned into a mega structure. As such, it contains many courtyards and intimate streets.

In this transformation lies the novelty of this project. The project is clear and rigorous, offering a very good urban elevation towards the arrival from the parking to the north as well as a very interesting small-scale series of elevations to the Place Lalla Yeddouna itself. The jury appreciated, in particular, the roof landscape with its potential for public use. However, regarding feasibility the project carries too many doubts, which led the jury to the decision not to award a prize to the project. -*Jury Comments*

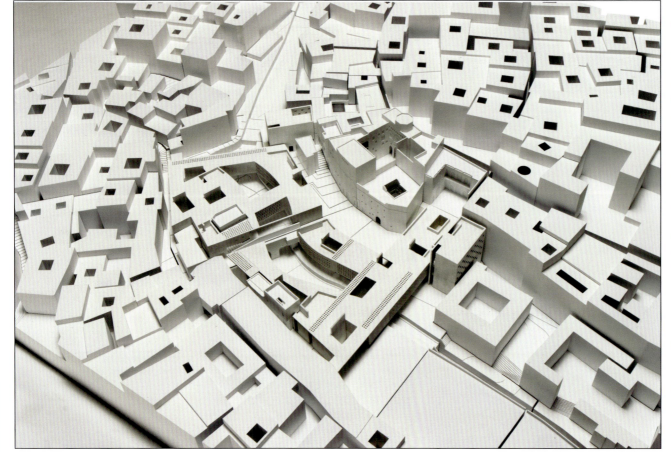

LEFT, ABOVE
Birdseye view of model
LEFT
Aerial view of site

Round 1 Finalist

Hanse Unit, Hamburg, Germany

第一轮入围奖

汉斯联合会（德国，汉堡）

这个方案的亮点在于巨大的穹顶连接了所有的新建筑，为游客提供了一个连续的散步长廊。这个吸引人的设计不仅增添了项目的现代氛围，而且还将它与千年古城完全隔绝了起来。

然而，评审们仍然怀疑这一基础概念是否可行。"改变活动路线"并让行人从地面到达屋顶可能会弱化典型的街区生活，引入一种让本地人感到突兀的生活方式。其次，这一理念的成功大部分取决于屋顶的质量。在图纸上，屋顶是一层薄薄的轻纱，但是设计师无法解释如何实现这种结构的屋顶。

本页
模型鸟瞰图

The quality of the scheme much depends on the idea of a large canopy connecting all the new buildings into a continuous roof-top promenade for tourists. Morphologically this presents an attractive idea as this large-scale element meets a contemporary mode of use but also isolates and frames one urban typical situation within this 1000-year-old city.

However, the jury is in doubt whether the fundamental idea will work. To "reroute" some of the activities and the associated pedestrian traffic from the ground onto the roof might weaken the typical street life and introduce a relationship to the city that –at least some of the local jurors– find alien. Secondly the success of the idea will largely also depend of the quality of the roof itself. It is shown on the drawings as a light veil but the authors fail to explain how they want to achieve this impression in the face of their own ambition to make this roof the carrier of some major technological components. Once wind and seismic requirements have been considered as well it might end up considerably more solid and hence much more present in a way that could not really be imagined to be in keeping with the world heritage of the site. Neither the relatively predictable floor plans nor the dutiful quotation of a collection of local materials reassured the jury that this might be a scheme that might excel beyond the one –admittedly at first sight seductive- scheme of the rooftop promenade. The idea of a decorative 'Garden of Eden' that cannot be entered 364 days a year located in the middle of a busy artisan's district where access to the water is essential seems to reinforce the reading of this site predominantly from a tourist's perspective.
-Jury Comments

THIS PAGE
Aerial studies of the model

Round 1 Finalist

Giorgio Ciarallo
Milan, Italy

第一轮入围奖

乔治奥·希亚拉罗（意大利，米兰）

评审们欣赏这个项目的勇气，它打造了一个
全新的积极城市规划策略。此外，评审们也
十分欣赏该项目的决心和采用传统建造技
术、材料和风格的愿望。然而，他们还是认
为这个项目过于夸张。

出于项目场地的地表特征，评审对新公共空
间的设计表示怀疑。河流四周和堤岸的设计
只可远观，不可使用。这样一来，人们能够
实际使用的公共空间只有一条狭窄的走道。
这与建筑师的原始愿望形成了冲突，是项目
的主要弱点。

本页
模型鸟瞰图

The jury appreciated the ambition of the project to create a new urban space as a very positive strategy. Furthermore, the jury appreciated the relatively high level of resolution of the proposal and the desire to engage and interpret traditional construction techniques, materials and architectural language. However it criticized the scheme for being rhetorical.

The jury remained unconvinced of the new public space given the compromised nature of its ground surface. The enclosure of the river and the surrounding embankments is actually designed as a space to look at rather than to be in. In turn, the public space that people would actually occupy is reduced to a mere corridor. This was seen to be in fundamental conflict with the architects stated intention to create a Gran-Fondouk and therefore regarded as a major weakness.

THIS PAGE
Aerial studies of the model

Round 1 Finalist

Arquivio Arquitectura
Madrid, Spain

第一轮入围奖

阿尔吉维奥建筑事务所（西班牙，马德里）

项目的清晰感深受好评。它没有重建现有的城市结构，而是提供了一个可以作为休闲区域的广场。人们可以在这里会面、交流和分享，或是庆祝传统节日。设计还在地下提供了大量商业活动空间，简单的网络将各个空间连接起来，更方便建造。项目凸显了菲兹的河畔之城特色，新建的广场在两个层面上都与河水相联系。然而，项目的质量取决于大型悬臂屋顶的实现。屋顶起到了基本的气候调节作用，并且对空间进行了划分。尽管评审对项目将雇用当地手工艺人十分满意，有限的预算和建造的复杂性之间产生了不可调节的矛盾。——评审评语

本页
展示材料和结构的模型鸟瞰图

The jury appreciated the clarity of the scheme, and the bold idea not to rebuild the existing structures and to offer a piazza instead that could serve as a recreation area and a place to meet, reflect and share, also for festivities like a traditional "Saha". The proposal also offers a generous space on the lower level to be used for commercial activities. A simple grid articulates the space and makes it easy to construct. This is a project that celebrates Fes as the city on the river, with the new piazza interacting with the water on two levels. However, the quality of the scheme fundamentally depends on the realization of the large hanging roofs as they provide essential climate moderation, and give scale as well as spatial definition. Given the available budget and the complexity of its construction this element was called into question even though the jury appreciated the intent to involve local craftsmanship in its production.
-Jury Comments

THIS PAGE
Aerial views of model illustrating materials and construction intents

NEW COURTYARD UPPER LEVEL

LAYERING THE PRESENT

Architecture is living heritage. The idea, style, structure and function reflects the social, technological and cultural identity of its time, and they are consistently recorded and remembered in our built environment. In this sense, the Old Seminary building must be considered as an accumulation of layers of important historical events, which means that the new addition to the School of Architecture must also be considered as an act of adding another layer of history. Therefore, the new addition is neither a resultant of literal imitation of formal features of the old nor a expressive element of contemporary architecture; it must rather be a outcome of reinterpretation of the existing structure's characteristics, and translation into a new form of architectural elements. As a result, the accumulated layers of the old and the new creates another interpretation as a whole.

First Place (1)

"Layering the Present"
Jihoon Kim
Sanghwan Park
JKSP architects
New York, NY

ABOVE
Courtyard perspective
LEFT
Site plan

一等奖（1）

"让现在层次化"
吉洪·金；三万·朴；JKSP建筑事
务所（纽约）

上图
庭院效果图
左图
场地规划图

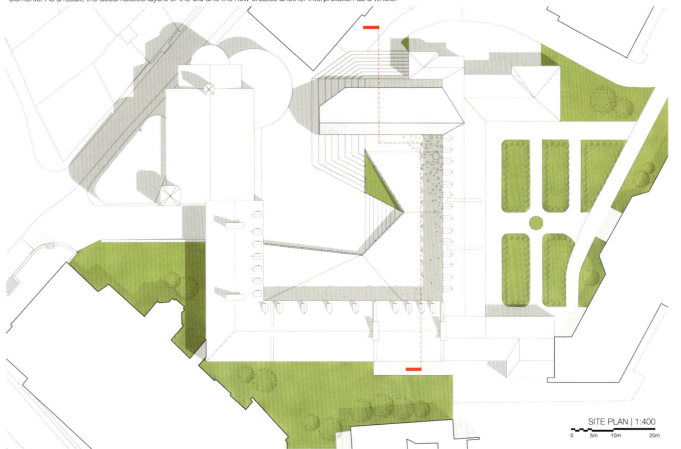

SITE PLAN | 1:400

0   5m   10m   20m

Honoring the Old, In with the New

Laval University School of Architecture "Emblematic Addition"

尊重过去，拥抱新纪元
拉瓦尔大学建筑学院 "象征性扩建"

Addressing an addition to a centuries-old seminary building in the heart of historic Québec—designated a UNESCO World Heritage Site— would represent a unique challenge to any architect. To probe the boundaries of this scenario, the Laval University School of Architecture, celebrating the 50th anniversary of its founding, sponsored a one-stage ideas competition, open to professionals and students alike for an "emblematic addition" to the heritage building where it resides.

The Site The existing Seminary Building is a U-shaped structure, surrounding an underutilized courtyard. This became a focal point for some of the schemes, as one of the winners stated: "The courtyard of the former seminary is today rarely used by the students of the school due to the lack of access and to the main floor functions—like the classrooms, being too specific." The competitors were also given maximum flexibility as to where the new structure might be located. According to the competition brief, "Its placement should reaffirm the entry hierarchy and the interior axes of circulation. Its volume should transform the building's exterior appearance, complementing it with a new symbolic icon.

The intervention, as a project embodying the values and dynamic of the School of Architecture, should meet the following objectives:
- Provide the School of Architecture with the gathering places that it currently lacks: a 350-seat amphitheater and a multi-use space of about 250 m2, which can serve as a space for exhibitions, studio critiques and an open forum.
- Reorient directionally and hierarchically the building ensemble in respect to the existing entrances and vertical circulation axes.
- Address the existing building, built several centuries ago, as a work in constant evolution—from its history as a local seminary to an internationally renowned school of architecture, with spaces that correspond to the pedagogical and research needs of today."

By staging a competition for a structural addition to its site, the Laval School of Architecture was furnished with a plethora of approaches to a spatial program. They are to be commended for their forward-looking design philosophy—seeking a contemporary solution to the frequent challenges posed by historic sites. There are many successful examples which Laval might follow; but solutions such as Richard Rogers' addition to the courts in Bordeaux illustrate how it can be done in a sensitive, but innovative fashion. -Ed

对魁北克中心区（世界文化遗产）一座百年教学楼进行扩建，对任何建筑师来说都是一个巨大的挑战。为了探索这个方案的界限，拉瓦尔大学建筑学院在成立50周年之际，组织了一次面向专业人士和学生的概念竞赛，为其所在的历史建筑进行"象征性扩建"。

场地　现有的教学楼呈U字形，以一个未充分利用的庭院为中心。这成为了一些参赛方案的焦点，一名优胜者称："由于缺乏入口以及主要楼层的功能太过明确，教学楼的庭院很少被学生们使用。"竞赛还对新结构的位置给予了很大的空间。根据竞赛提纲，"它的位置应该巩固教学楼入口和内部空间轴线。它应该改变建筑的外观，使其具有象征意义。"

项目设计将展现建筑学院的价值和活力，需要实现以下目标：
–为建筑学院提供一个其所欠缺的集会地点：一个可容纳350人的露天剧院和一个约205平方米的多功能空间。这个多功能空间按可用作展览、工作室讨论区和开放式论坛。
–在尊重现有入口和垂直交通中轴的基础上，调整建筑整体的定向和层次。
–将拥有百年历史的原有建筑看做一个不断进化的过程——从本地神学院发展成为一座闻名海外的建筑学院。打造适应当前教学研究需求的全新空间。

通过举办这次建筑竞赛，拉瓦尔建筑学院获得了大量空间设计方案。这些方案前卫的设计哲学以及试图在历史场地中增加现代设计的探索值得赞赏。拉瓦尔学院拥有许多能够效仿的成功范例；但是，类似理查德·罗杰斯在波尔多的新建项目的方案展示了如何以适当而创新的方式进行新建工程。——编者

NEW COURTYARD LOWER

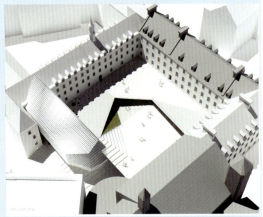

PERFORATED FACADE PATTERN

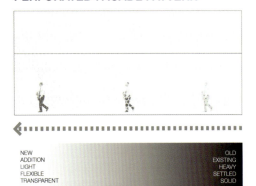

NEW   OLD
ADDITION   EXISTING
LIGHT   HEAVY
FLEXIBLE   SETTLED
TRANSPARENT   SOLID

CIRCULATION & NEW PROGRAMS

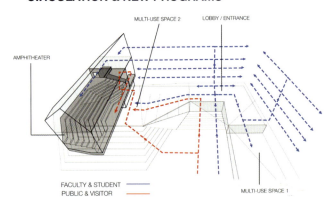

MULTI-USE SPACE 2   LOBBY / ENTRANCE

AMPHITHEATER

FACULTY & STUDENT
PUBLIC & VISITOR

MULTI-USE SPACE 1

AMPHITHEATER ADDITION   WING OF PROCURE

UNFOLDED SECTION | 1:250

0   2m   5m   10m

THEATER NIGHT VIEW

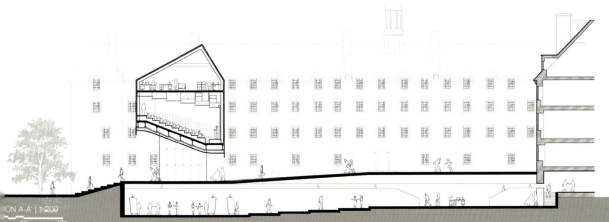

ON A-A' | 1:200
5m   10m

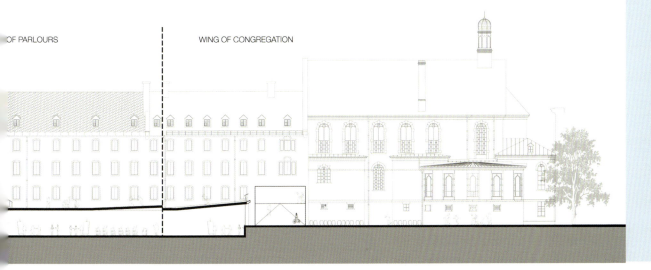

OF PARLOURS   WING OF CONGREGATION

First Place (1)

"Layering the Present"
Jihoon Kim
Sanghwan Park
JKSP architects
New York, NY

The authors of this scheme utilized the interior courtyard to its utmost in developing their program. According to their statement, "the Old Seminary building must be considered as an accumulation of layers of important historical events, which means that the new addition to the School of Architecture must also be considered as an act of adding another layer of history. Therefore, the new addition is neither a resultant of literal imitation of formal features of the old nor an expressive element of contemporary architecture; it must rather be a outcome of reinterpretation of the existing structure's characteristics, and translation into a new form of architectural elements. As a result, the accumulated layers of the old and the new create another interpretation as a whole." -Ed

LEFT	左图
Competition boards	竞赛展示板
FAR LEFT, ABOVE	左上
Birdseye view	鸟瞰图
FAR LEFT, BELOW	左下
Interior view	室内景象

一等奖（1）

"让现在层次化"
吉洪·金；三万·朴；JKSP建筑事务所（纽约）

该项目的设计师充分利用了内部庭院。根据他们的表述，"旧教学楼被看成了多层重要历史事件的累积，而新增的建筑学院结构也必须是一层新的历史。因此，扩建结构既不能是对过去的模仿，又不能是过于表现现代建筑元素；它必须是对现有建筑特色的重新解读，展现全新的建筑元素。这样一来，新旧层次的叠加共同打造了一个整体。"——编者

First Place (2) 一等奖（2）

"直轴"

"Direct Axis"
Omar Aljebouri
Avery Guthrie
Steve Socha
Toronto, Canada

奥马尔·阿尔及伯里；艾弗里·格思里；史蒂
夫·苏哈（加拿大，多伦多）

下图
2号竞赛展示板的说明图
右图
1号竞赛展示板

BELOW
Illusrations from competition board 2
RIGHT
Competition board 1

Direct Axis

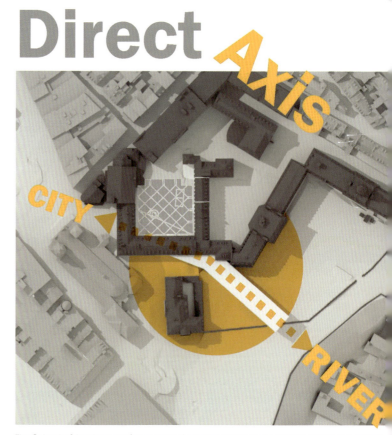

CITY
RIVER

The School of Architecture forms part of a larger complex of buildings that have served, historically, as the Seminary of Quebec and the original home of Laval University. The tradition of building on the site has been one of addition and adaptation; new growth is accommodated through the accumulation of axial wings, with form and programmatic function evolving to suit the needs of successive ecclesiastical and pedagogical eras. A intervention on this site must thus treat the context as a dynamic process, rather than a group of static artifacts. This project aims to take part in this process through the addit of the next axis in the Old Seminary complex.

The project acknowledges the fact that each addition to the seminary site both creates new interior volumes and relationships, but also defines new outdoor spaces. While this practice has created the school's most prominent public space, the central courtyard, i has also resulted in an often unintelligible circulation system and a condition in which the architecture school is largely cut off from its wider context. The creation of a new ax establishes a clear and decongested circulation path that both satisfies the architectur school's contemporary spatial needs as well as reestablishes a strong visual link to the city and river valley.

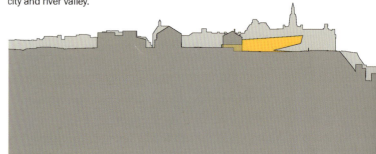

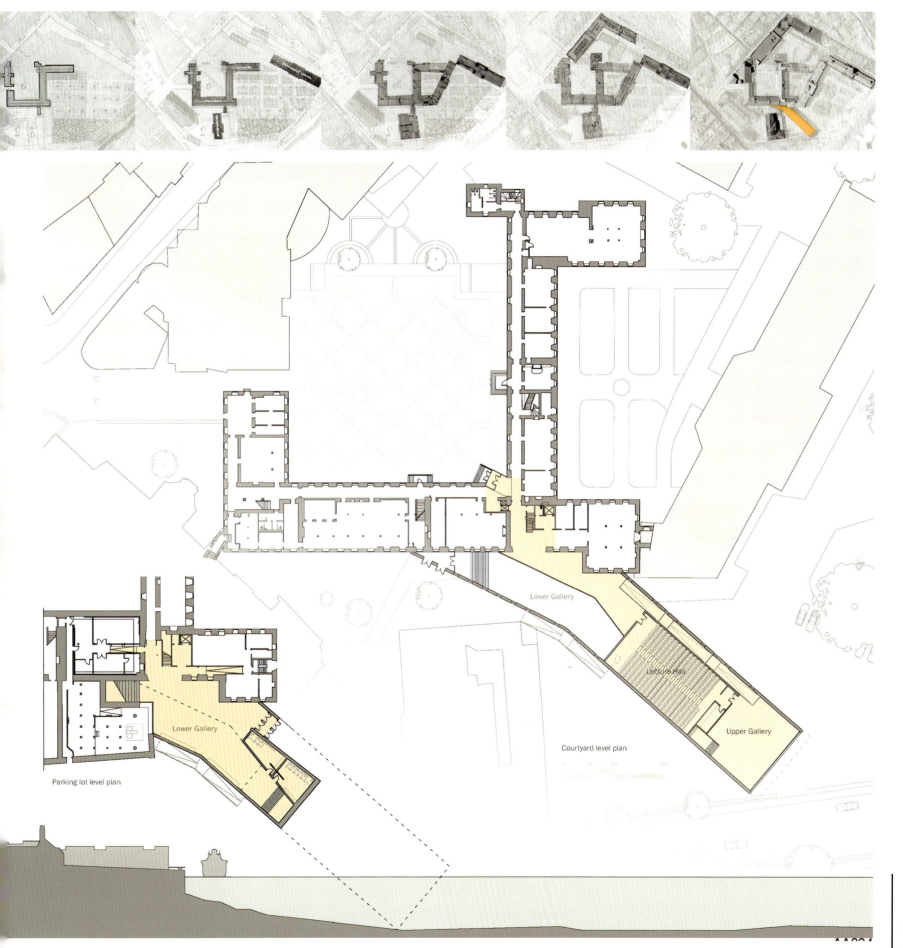

Lower Gallery

Courtyard level plan

Lecture Hall

Upper Gallery

Lower Gallery

Parking lot level plan

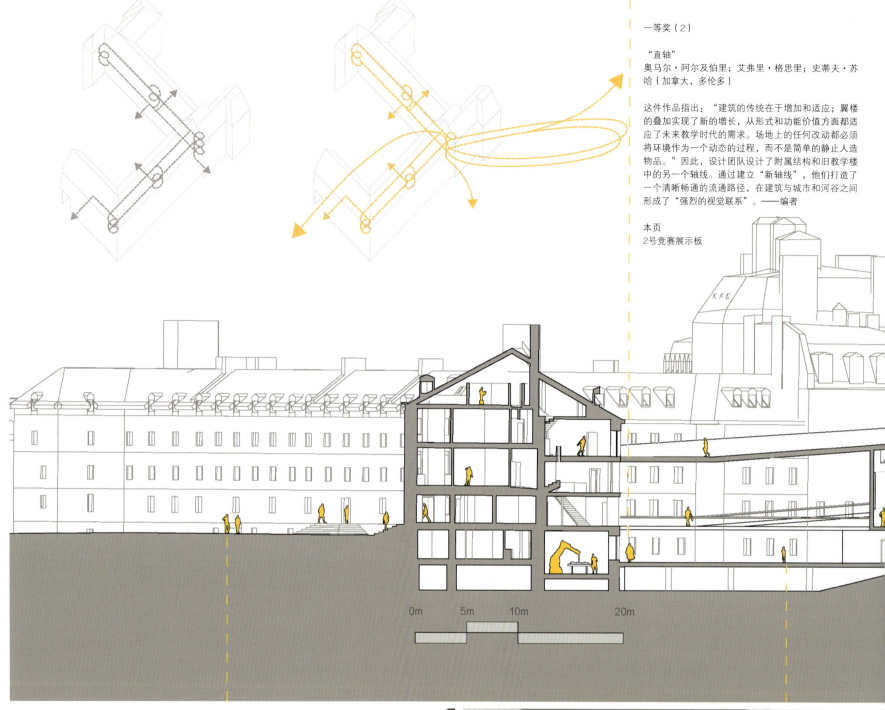

一等奖（2）

"直轴"
奥马尔·阿尔及伯里；艾弗里·格思里；史蒂夫·苏哈（加拿大，多伦多）

这件作品指出："建筑的传统在于增加和适应；翼楼的叠加实现了新的增长，从形式和功能价值方面都适应了未来教学时代的需求。场地上的任何改动都必须将环境作为一个动态的过程，而不是简单的静止人造物品。"因此，设计团队设计了附属结构和旧教学楼中的另一个轴线。通过建立"新轴线"，他们打造了一个清晰畅通的流通路径，在建筑与城市和河谷之间形成了"强烈的视觉联系"。——编者

本页
2号竞赛展示板

0m   5m   10m   20m

First Place (2)

"Direct Axis"
Omar Aljebouri, Avery Guthrie, Steve Socha
Toronto, Canada

This entry points out that "the tradition of building on the site has been one of addition and adaptation; new growth is accommodated through the accumulation of axial wings, with form and programmatic function evolving to suit the needs of successive ecclesiastical and pedagogical eras. Any intervention on this site must thus treat the context as a dynamic process, rather than a group of static artifact." Thus, this team suggested an addition and another axis in the Old Seminary complex. By establishing a "new axis," a clear and decongested circulation path is created, also resulting in a "strong visual link" to the city and river valley. -Ed

THIS PAGE
Competition board 2

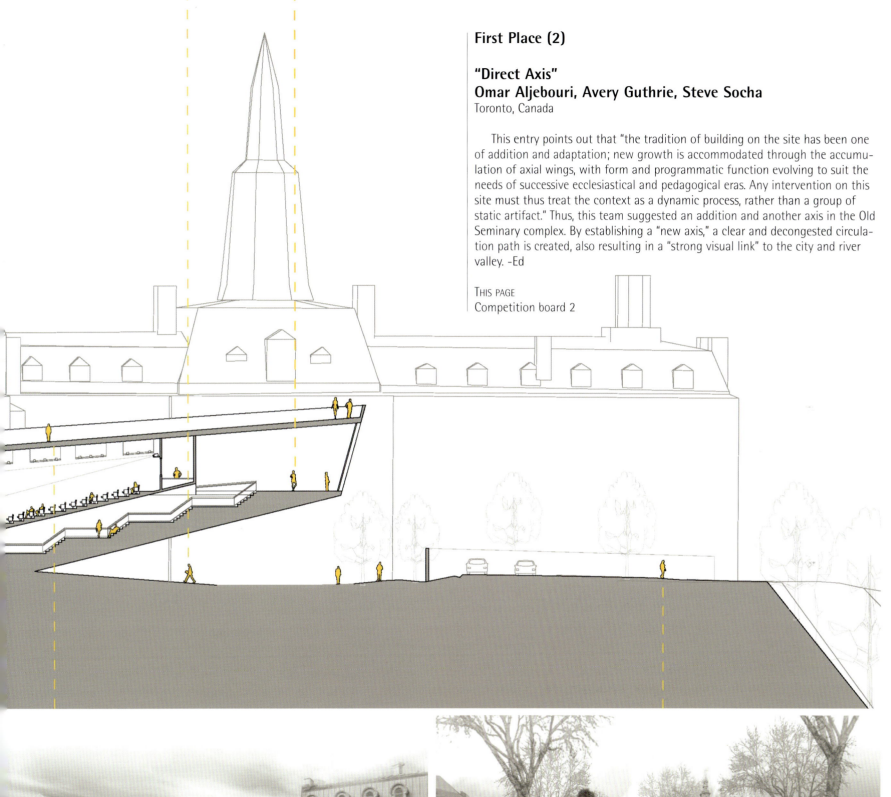

Se déplacer vers les gens

Sortir l'architecture de ses murs

Patiner

Séquestration

Aménagement libre =
Exposition de structures
Pique-nique
Étude en plein-air

Surfaces perméat.

First Place (3)

"Shed"
Catherine Houle
Marianne Lapalme
Vanessa Poirier
Laval University
Québec, Canada

THIS PAGE AND OPPOSITE
Illusrations from competition
board 1

一等奖（3）

"分水岭"
凯瑟琳·休尔；玛丽安·拉帕尔姆；
瓦内萨·波伊利尔（加拿大，魁北
克，拉瓦尔大学）

本页和对页
1号竞赛展示板示意图

ÉLÉVATION SUD-EST   0m ▬▬▬ 2m

SHED

Système Hybride d'Expérimentation et de Diffusion

Intégration de la gestion lumière-thermique-eau

Ancré vs Déposé

Empreinte minimale

Expérimenter avec la matière

Projection de films

ORIENTATIONS ENVIRONNEMENTALES, PRINCIPES D'ARCHITECTURE BIOCLIMATIQUE

ÉNERGIE D'OPÉRATION
Chauffage solaire passif
Masse thermique
Éclairage naturel
Récupération des eaux pluviales
Ventilation transversale
Effet de cheminée
Refroidissement passif

SÉQUESTRATION DE CARBONE PAR LES MATÉRIAUX
Structure minimale en bois

TRANSPORT
Espace couvert pour stationnement vélos
Proximité des transports en commun
Encourage la mobilité vers les infrastructures existantes

OCCUPATION DU SITE
Implantation minimale, respectueuse du site
Perturbation minimale du site
Réhabilitation d'un bâtiment existant

ÉNERGIE PRODUITE
Cellules photovoltaïques

QUELS SONT NOS VÉRITABLES BESOINS? QU'EST-IL NÉCESSAIRE DE CONSTRUIRE?

Requestionner les besoins est la première étape afin d'établir un programme d'ajout manifeste en véritable accord avec les priorités actuelles de l'École d'Architecture et les enjeux du lieu.

Partant de l'idée que 80 % de l'énergie consommée par un bâtiment sur son cycle de vie est liée à sa période d'opération, comment faire plus avec moins ?

Qu'est-il nécessaire de construire au fond ?

Quand des laboratoires d'exception sont disponibles au pavillon Kruger et que des salles parfaites pour la tenue de conférences sont situées à distance de marche, pourquoi construire alors que les espaces existent déjà ?

L'utilisation des infrastructures présentes à travers la ville permet d'envisager une diminution significative du programme suggéré.

Les meilleurs mètres carrés sont ceux qu'on ne construit pas.

Et quant est-il de l'étudiant ?
À l'ère du numérique, comment apprend-on à faire de l'architecture ?

Peu importe les logiciels, les rendus numériques ou les simulations virtuelles... au bout du compte, l'architecture ne sera-t-elle pas perpétuellement confrontée à la matière ?

L'étudiant doit donc explorer et tester, mais aussi ressentir et expérimenter.

Dans l'optique d'une intervention minimale, l'espace dont l'école d'architecture a réellement besoin n'est-il pas un vaste espace destiné à la construction de structures, mais permettant également les usages les plus divers ?

Réduire la superficie construite, diminuer la perturbation du site, utiliser des matériaux renouvelables et séquestrant le carbone, gérer les éléments lumière et eau à travers un système simple, l'ensemble de ces principes ne mènent-t-il pas à la carboneutralité ?

Dans l'optique de favoriser la mobilité, pourquoi tenter d'attirer la population vers soi lorsqu'on peut se déplacer vers elle ? Favoriser la mobilité permet d'aller au devant, de véritablement atteindre un public cible dans l'optique de démocratiser l'architecture.

Investir la ville, intégrer les lieux publics, se déplacer vers les gens, se fondre aux diverses manifestations artistiques et culturelles... voilà la voie à suivre pour démocratiser l'architecture !

Il faut sortir l'architecture de ses murs !

La SHED est avant tout un espace de construction de structures permettant l'exploration architecturale à travers la matérialisation des idées dans l'optique de déplacer et diffuser l'architecture.

En ville ou à la campagne, sous une bretelle d'autoroute bruyante ou dans un champs venteux, le territoire tout entier est un laboratoire d'architecture. Il faut se donner les moyens de s'y déplacer et de rendre tangible les imaginations intangibles.

ÉCOLE D'ARCHITECTURE

vers Le Cercle

vers pavillon Kruger

vers Musée de la Civilisation

AA186

PLAN D'IMPLANTATION　● PROXIMITÉ DES TRANSPORTS EN COMMUN
0m　　　50m　　　● FAVORISER LA MOBILITÉ VERS LES INFRASTRUCTURES EN PLACE

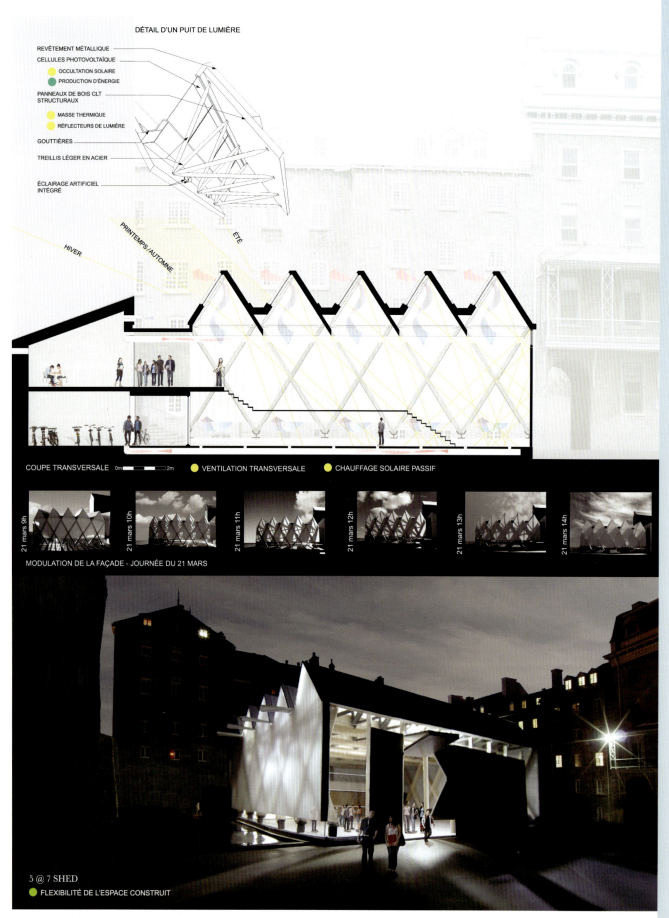

DÉTAIL D'UN PUIT DE LUMIÈRE

REVÊTEMENT MÉTALLIQUE
CELLULES PHOTOVOLTAÏQUE
　　OCCULTATION SOLAIRE
　　PRODUCTION D'ÉNERGIE
PANNEAUX DE BOIS CLT
STRUCTURAUX
　　MASSE THERMIQUE
　　RÉFLECTEURS DE LUMIÈRE
GOUTTIÈRES
TREILLIS LÉGER EN ACIER
ÉCLAIRAGE ARTIFICIEL
INTÉGRÉ

HIVER　　PRINTEMPS/AUTOMNE　　ÉTÉ

COUPE TRANSVERSALE　0m ▭ 2m　● VENTILATION TRANSVERSALE　● CHAUFFAGE SOLAIRE PASSIF

21 mars 9h　21 mars 10h　21 mars 11h　21 mars 12h　21 mars 13h　21 mars 14h

MODULATION DE LA FAÇADE - JOURNÉE DU 21 MARS

5 @ 7 SHED
● FLEXIBILITÉ DE L'ESPACE CONSTRUIT

First Place (3)

"Shed"
Catherine Houle
Marianne Lapalme
Vanessa Poirier
Laval University
Québec, Canada

The giant "shed" is at the same time an icon at the edge of the Seminary and a giant, flexible space for exhibits and other functions. Strangely, although not intended as a temporary structure, this proposal reminds one of the early Quonset hut structures used by some schools of architecture before permanent buildings were available (Ball State University), or the conversion of a gymnasium at Catholic University in Washington, D.C. As such, students are hardly pre-programmed or intimidated by the architecture and might even produce better design as a result. Even if this new added space is not for studios, symbolically it can send an interesting message. -Ed

THIS PAGE AND OPPOSITE
Competition board illustrations

一等奖（3）

"小屋"
凯瑟琳·休尔；玛丽安·拉帕尔姆；瓦内萨·波伊利尔（加拿大，魁北克，拉瓦尔大学）

巨大的"小屋"同时也是教学楼边缘的重要标志，提供了巨大而灵活的多功能展览空间。奇怪的是，尽管这不是一个临时结构，项目还是让人想起一些建筑学院建造的半圆拱形活动房屋（例如保尔州立大学）或者是华盛顿天主教大学的体育馆。如此一来，学生们很难对建筑进行模仿，甚至能够建造更好的设计。尽管这个新建空间不是为工作室而设计的，它能够发送出有趣的信息。——编者

本页和对页
竞赛展示板示意图

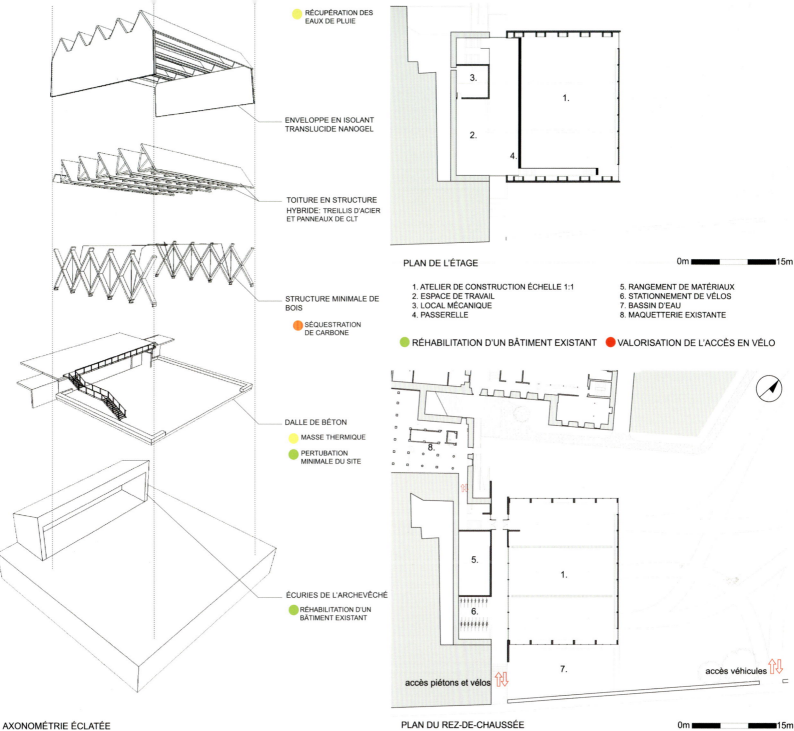

○ RÉCUPÉRATION DES
EAUX DE PLUIE

ENVELOPPE EN ISOLANT
TRANSLUCIDE NANOGEL

TOITURE EN STRUCTURE
HYBRIDE: TREILLIS D'ACIER
ET PANNEAUX DE CLT

STRUCTURE MINIMALE DE
BOIS

● SÉQUESTRATION
DE CARBONE

DALLE DE BÉTON

● MASSE THERMIQUE

● PERTUBATION
MINIMALE DU SITE

ÉCURIES DE L'ARCHEVÊCHÉ

● RÉHABILITATION D'UN
BÂTIMENT EXISTANT

AXONOMÉTRIE ÉCLATÉE

PLAN DE L'ÉTAGE

0m ▬▬▬ 15m

1. ATELIER DE CONSTRUCTION ÉCHELLE 1:1   5. RANGEMENT DE MATÉRIAUX
2. ESPACE DE TRAVAIL   6. STATIONNEMENT DE VÉLOS
3. LOCAL MÉCANIQUE   7. BASSIN D'EAU
4. PASSERELLE   8. MAQUETTERIE EXISTANTE

● RÉHABILITATION D'UN BÂTIMENT EXISTANT   ● VALORISATION DE L'ACCÈS EN VÉLO

accès piétons et vélos ⇅   accès véhicules ⇅

PLAN DU REZ-DE-CHAUSSÉE

0m ▬▬▬ 15m

● SÉQUESTRATION DE CARBONE PAR LES MATÉRIAUX

● LUMIÈRE NATURELLE ABONDANTE

AA186

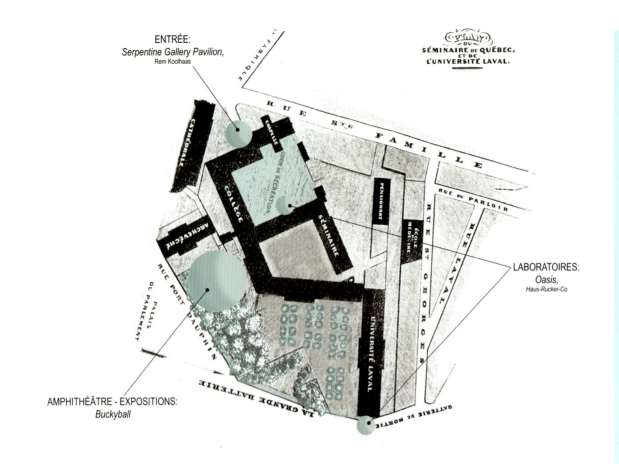

ENTRÉE:
Serpentine Gallery Pavilion,
Rem Koolhaas

SÉMINAIRE DE QUÉBEC, ET DE L'UNIVERSITÉ LAVAL.

LABORATOIRES:
Oasis,
Haus-Rucker-Co

AMPHITHÉÂTRE - EXPOSITIONS:
Buckyball

Honorable Mention

"spheres de references"
Éric Boucher
Élisabeth Bouchard
Montréal, Canada

According to the authors, "the sphere was chosen here for its contrast with the classic rigidity of the Seminary's classic architecture. The literal transparency of the sphere allows the appreciation of the existing structure (Hearst Tower as reference). This pure form repeated at multiple scales and places creates a streetscape celebrating the school within the city. The circulation forms a network, framed by the spheres. The typology chosen is only one part; the idea is to make reference to the architecture by the architecture itself." Here, contrast alone is presented as the solution. The problem here could well be content, as iconic ideas by themselves may be a fad in time. -Ed

THIS PAGE AND OPPOSITE
Competition board illustrations

荣誉奖

"参考球体"
埃里克·鲍彻；伊利莎白·布夏尔（加拿大，蒙特利尔）

根据设计师所说："我们之所以选用球体，是因为它与教学楼的经典造型形成了鲜明对比。球体的透明度让人们能够欣赏现有结构（以赫斯特大厦为参照物）。这个纯粹的造型以多重规模不断重复，形成了街景。内部流通路径形成了网络，以球体为边框装饰。设计概念以建筑参考建筑。"对比就是设计的重点。项目的问题在于在不久的将来，它很容易变得过时。——编者

本页和对页
竞赛展示板示意图

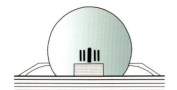

Maison idéale,
Claude-Nicolas Ledoux, 1770.

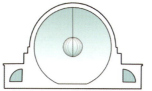

Cénotaphe pour Sir Isaac Newton,
Etienne-Louis Boullée, 1784.

Globe tower, Coney Island,
Samuel Friede, 1906.

Dome géodésique,
Buckminster Fuller, US patent 1954.

the Environment bubble,
François Dallegret, 1965.

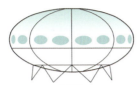

Futuro,
Matti Suuronen, 1968.

Die wolke,
Coop Himmelblau, 1970.

Oase no.7,
Haus-Rucker-Co, 1972.

Serpentine gallery pavilion,
Rem Koolhaas, 2006.

RAK convention center,
OMA, 2007.

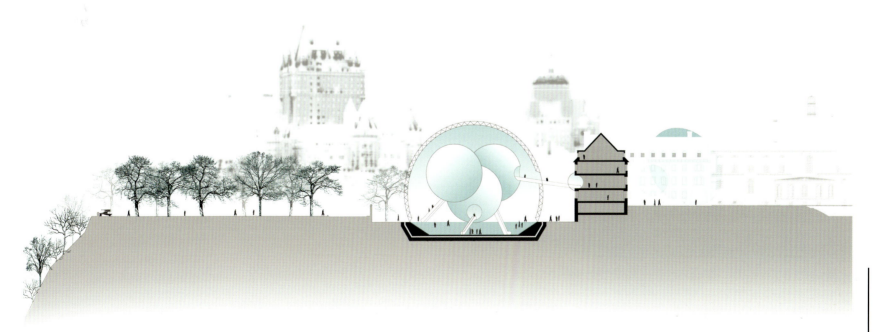

UNIVERSITE LAVAL
AJOUT MANIFESTE - PROJETER LE PATRIMOINE

Le vieux Séminaire de Québec fait partie d'un site historique fabriqué d'un assemblage de bâtiments qui ensemble constituent le monument historique dans lequel s'intègre l'école d'architecture.

A l'occasion du cinquantième anniversaire de l'école, il était temps pour celle-ci de faire signe et de manifester sa présence vers l'extérieur par une intervention qui s'inscrit dans le droit fil de la fabrication du site.

Cette intervention, pour l'école d'architecture, se doit naturellement d'être contemporaine, or, au vingt-et-unième siècle, on ne peut plus penser une telle intervention en termes d'arrachement, de coupure, de rupture ou de simple adjonction.

Notre ajout se pense d'abord comme une ouverture du site au monde, un outil de représentation, de dispenseur de parole et de savoir; un concentreur de communication.

C'est un condensateur d'énergie, à la fois capteur et diffuseur; sa peau reflète l'histoire puisque le monument s'y reflète de manière mouvante et changeante; elle reflète les passants aussi, les visiteurs, les chercheurs, les étudiants, pour les confondre avec l'histoire et les projeter dans l'avenir.

Un condensateur de savoir, d'histoire et des échanges, qui irradie et nous renvoie le ciel et les fabuleuses lumières de Québec.

COUPE TRANSVERSALE

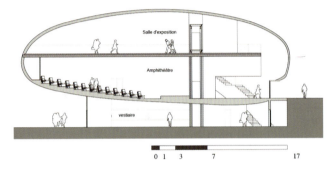

NIVEAU BAS

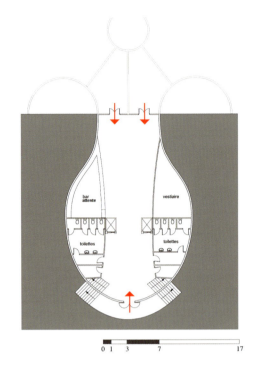

NIVEAU REZ DE CHA

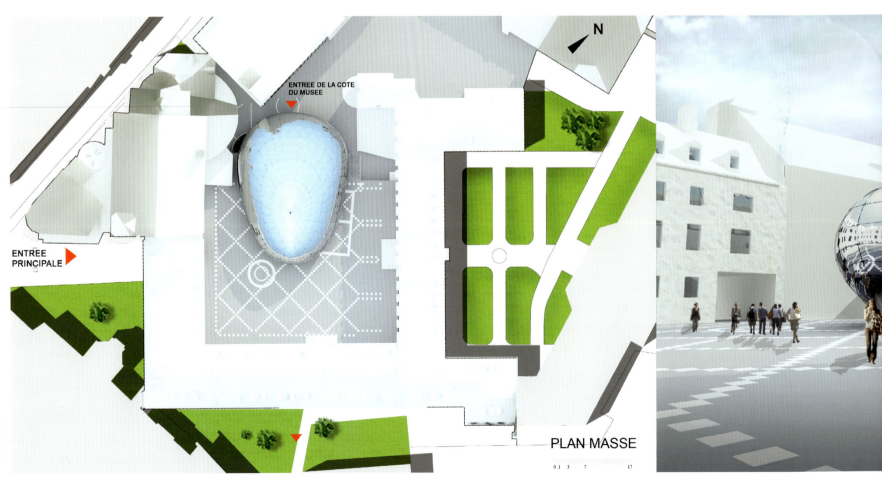

PLAN MASSE

NIVEAU +1

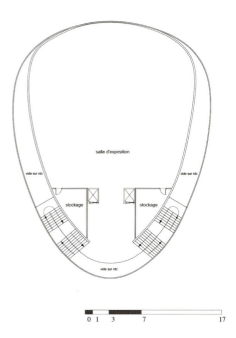

salle d'exposition

vide sur rdc

vide sur rdc

stockage stockage

vide sur rdc

0 1 3 7 17

Honorable Mention

Odile Decq
Paris, France

This single, iconic object is intended to indicate by its presence to the outside world that this is something other than a religious institution, with emphasis on the 21st century. It is a means of representation to the outside world. Its shiny skin simultaneously reflects both history of the monument and passersby. Here one cannot ignore the similarity of this structure with Anish Kapoor's popular 'bean' in Chicago's Millennium Park, the exception being that here it serves an additional purpose. -Ed

荣誉奖

奥蒂勒·戴克（法国，巴黎）

这个独立的地标性建筑试图向外界展示：自己并非仅仅是一座宗教学院，还具有21世纪气息。项目是面向外界的展览。闪闪发亮的建筑表皮倒映出历史建筑和行人的身影。除了功能不同之外，人们无法忽视项目与安尼诗·卡普尔所设计的芝加哥千禧公园"豆荚"的相似性。——编者

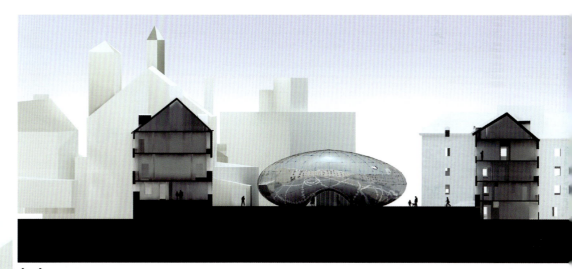

ÉLÉVATION FAÇADE SUD EST

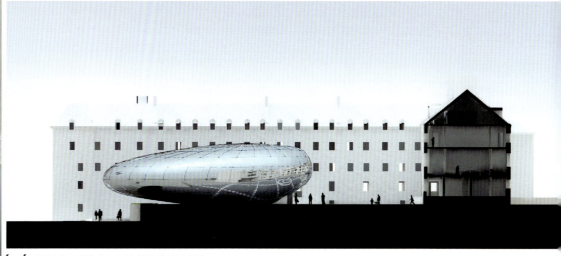

ÉLÉVATION FAÇADE SD OUEST

N° AA 154

EPLAND QUÉBEC *clic!* *clic!*
RIMOINE MONDIAL

ctionne pas. La conservation irréfléchie tue
désertées par ses habitants et la population
st du toc! Il est temps de passer aux choses
sérieuses.

une ville figée, nous ne vivrons pas dans le
. Exit la fixation dans le temps, fini le botox
, nous allons vivre la ville que nous aimons.

ulation de strates : nous voulons brasser les
nétisser les époques. **Vivre le patrimoine.**

rd. L'Université Laval revient au berceau de
ennent vie, la vieille capitale vibre, Québec
u début des années 1960, les signes vitaux
retour de l'Université Laval dans ses murs.

naire. Un étage commun qui relie tous les
s l'ancien. La strate est lieu d'échange, de
e en commun et de partage de ressources.

ANIFESTE

Honorable Mention

SHLACK!!!
Julien Beauchamp
Romy Brosseau
Émilie Gagné-Loranger
Alexandre Hamlyn
Laval University, Québec

The authors point out that the existing building doesn't work. Whereas the existing pathways footpaths link all the programs now, an added level to all the buildings can serve as a common contemporary space, which can also serve as a place for free exchange of ideas. This "de-cloistering" of the site can promote interrelationships between students and faculty. This solution is somewhat reminiscent of the old Italian courtyards with the surrounding balconies, albeit enclosed because of the winter weather. In any case, this is a way to connect outside the stolid walls of the existing structure. -Ed

荣誉奖

沙拉克！！！
朱利安·比彻姆；罗密·布罗索；艾米利·加涅-罗兰果尔；亚历山大·哈姆林（魁北克，拉瓦尔大学）

设计师指出原有建筑已经不起作用。但是他们利用原有的走道连接了各个功能区。每座建筑上方的加层提供了现代公共空间，供人们交流沟通。设计瓦解了场地的回廊特征，促进了师生之间的互动。这一设计让人想起了旧式意大利庭院与周围的阳台。设计将原有结构冰冷的墙壁与外部空间联系了起来。——编者

TADAM !!

CAMPUS DE DESIGN

Arts Visuels · Design Graphique · Design Industriel · Architecture · Aménagement · Salle de Critiques · Amphithéâtre

Laboratoire Informatique · Salles d'étude · Café Lounge · Salle d'Impression · Local de sérigraphie · Local de peinture · Maquetterie · Cafétéria · Salles d'Exposition · Magasin Scolaire · Associations Étudiantes · Amphithéâtre

Circulations verticales existantes

Toutes les disciples de design sont rassemblées pour former un campus de design. En s'inspirant des passerelles ayant déjà existées, les différents domaines sont connectés les uns aux autres.

L'étage ajouté à tous les bâtiments du campus de design sert de lieu commun. C'est espace contemporain devient le lieu des échanges et des partages qui enrichiront chacune des disciples.

Le décloisonnement des pavillons à cet étage favorise le flânage et l'errance à travers tout le campus, engendrant ainsi des rencontres inatendues.

L'entrée actuelle de l'école d'architecture s'avère comme inadaptée par ses dimensions, sa localité et ses expressions pour recevoir un visiteur.
Le projet propose donc une entrée qui se fait dans la cour. Elle devient donc le **premier lieu d'accueil**.
Une fois passé la porte cochère, le visiteur est guidé par une pente qui le dirige vers la nouvelle entrée principale ensoleillée par le soleil du sud. Les locaux de l'entrée, spacieux et lumineux, correspondent aux besoins contemporains et mènent directement vers les lieux les plus publiques c'est à dire à l'auditorium et à la salle d'exposition.
Aussi une liaison directe avec la circulation verticale est facilement accessible.
Les escaliers de l'**auditorium profitent du dénivellement** du terrain.
La cour de l'ancien séminaire est aujourd'hui rarement utilisée

par les étudiants de l'école; faute de manque d'accès à cette dernière et en raison de fonctions trop spécifiques du rez-de-chaussée actuel comme des salles de cours.
Les caractéristiques typiques d'un séminaire reposent entre autre sur la typologie de la cour fermée.
Le projet propose donc de venir **fermer la cour**, tout en s'appliquant sur les anciens plans du séminaire où cette fermeture était prévue. Elle garantie une **réappropriation de la cour** par les étudiants. Elle est englobée par des fonctions publiques et semi-publiques qui font de la cour et du nouveau rez-de-chaussée un **grand lieu de rassemblement**. La cour est entièrement entourée par les parois vitrées du nouvel emblème qui renforcent les vues et l'accessibilité de celle-ci.
Le projet s'implante comme une ligne légère tout au long de la cour.

8

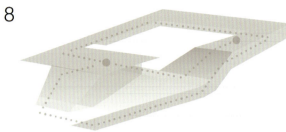

••••••• la circulation se ferme

atelier de bois cafétéria	FONCTIONS SEMI-PUBLIQUES
entrée auditorium	
salle d'exposition	FONCTIONS PUBLIQUES

PLAN DU SITE | échelle 1.1000

Idendification **AA014**

"身份"
吉尔·本茨（卢森堡）

AJOUT MANIFESTE

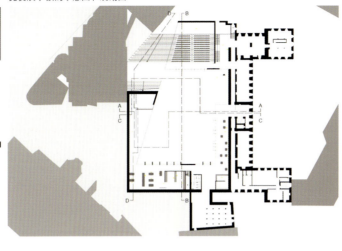

Citation

"Idendification"
Jil Bentz
Luxembourg

This is one of the more remarkable entries for its use of the topographical features of the site. This is part architecture, part landscape, making it an interesting fit. -Ed

THIS PAGE AND OPPOSITE
Illustrations and plans from the competition boards

CONCEPT

1 une partie de la cour est baissée d'un niveau pour créer un nouveau rez-de-chaussée

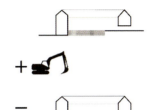

2 entrée dans la cour I chemin défini pour accéder à l'entrée du bâtiment universitaire

3 **entrée** fourni un accueil approprié à une université I ⬤ relation avec circulation vertical

4 l'emplacement de l'**auditorium** ferme la cour du séminaire

l'auditorium profite du dénivellement

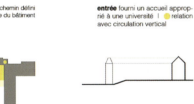

5 le **café** touche les anciens mures uniquement par son toit

le volume du café crée une sorte de base à l'aile de la procure pour garder les anciennes proportions
le café crée également le lien avec les jardins et le stationnement

6

7

Idendification AA014

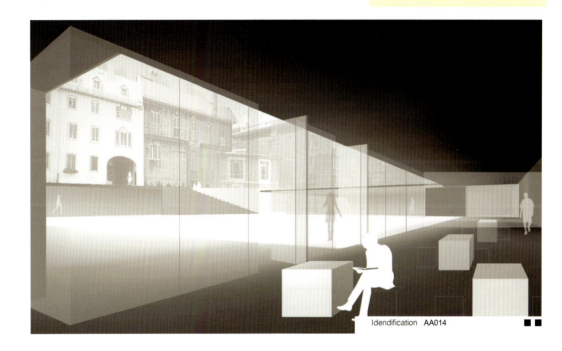

COUPE A.A

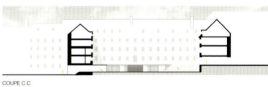

COUPE C.C

COUPE B.B

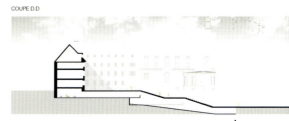

COUPE D.D

ÉCHOS

LA VIE D'UN BÂTIMENT EST PONCTUÉE D'ÉVÉNEMENTS QUI ONT LEURS ÉCHOS DANS LE TEMPS.

CES ÉVÉNEMENTS MARQUENT LE BÂTI ; LEURS CONSÉQUENCES S'ADDITIONNENT, SE SUPERPOSENT ET FRAPPENT À L'UNISSON NOTRE PERCEPTION.

CES ÉCHOS DU PASSÉ ONT FAÇONNÉ LE BÂTI TEL QUE NOUS LE CONNAISSONS AUJOURD'HUI ET FORMENT UN TOUT. C'EST À CELA QUE NOUS AJOUTONS UN TRAVAIL QUI AURA SA RÉSONANCE PROPRE AU SEIN DE L'ENSEMBLE.

PÉDAGOGIE

LE MOT "PÉDAGOGIE" DÉRIVE DES TERMES GRECS SIGNIFIANT "ENFANT" ET "ACCOMPAGNER". DANS L'ANTIQUITÉ, LE PÉDAGOGUE, CHAQUE JOUR, ACCOMPAGNAIT L'ENFANT À L'ÉCOLE ET PORTAIT SES AFFAIRES TOUT EN LUI FAISANT RÉCITER SES LEÇONS ET FAIRE SES DEVOIRS.

DEPUIS SES DÉBUTS, LE SÉMINAIRE PARTICIPE À LA TRANSMISSION DU SAVOIR.

IL EST INTÉRESSANT DE NOTER QUE LE PROCESSUS D'APPRENTISSAGE PASSE FORCÉMENT PAR UNE RÉINTERPRÉTATION TRÈS PERSONNELLE DE LA MATIÈRE PAR L'ÉTUDIANT (TOUT COMME LA MATIÈRE EST INTERPRÉTÉE ET TRANSMISE DE MANIÈRE TRÈS PERSONNELLE PAR L'ENSEIGNANT). MÊME SI L'ESSENCE DEMEURE INTACTE, LES PERCEPTIONS DIFFÈRENT AU GRÉ DU TEMPS ET DES GENS. CETTE DIVERSITÉ D'INTERPRÉTATION EST GARANTE D'UNE SAINE ÉVOLUTION.

TYPOLOGIE

LA DÉCOUVERTE DE CET ENSEMBLE AILES ET COURS PROPOSE UNE SUCCESSION DE PLANS : BÂTI SUR COUR SUR BÂTI SUR COUR SUR BÂTI...

LES AILES DU BÂTIMENT PRENNENT LA FORME DE GRANDS MONOLITHES BLANCS, FORTEMENT ANCRÉS DANS LE CAP ET DANS L'HISTOIRE.

POSITION

LES ÉCHOS S'ADDITIONNENT, LES LANGAGES SE SUPERPOSENT ET NOS SENS RÉINTERPRÈTENT...

QUELLE EST L'ESSENCE ARCHITECTURALE DU SÉMINAIRE ? COMMENT NOS SENS LA PERÇOIVENTELLE ?

L'AJOUT MANIFESTE EST UN ÉCHO DU VIEUX SÉMINAIRE : IL SE VEUT L'ÉCHO D'UNE SUITE D'ÉCHOS. LE REFLET DU MIROIR DANS UN MIROIR : UNE RÉFLEXION INFINIE, UN COMMENTAIRE SANS FIN.

PLUS PRÉCISÉMENT, C'EST UNE RÉPONSE À L'ENTRÉE EXISTANTE DE L'ÉCOLE D'ARCHITECTURE : UNE RÉPONSE QUI SEMBLE JAILLIR DE LOIN CAR LA VOIE EST DÉFORMÉE... MAIS LE MESSAGE RESTE CLAIR.

DÉCONS
RÉAFFIR

" DE
LAISSER
DIS BIEN
L'AUTRE
VENTE
L'INVENT
NE PE
TER À
DÉSTABI
DE FORC
PASSAGE

L'AUTRE

10H00   12H00   14H00

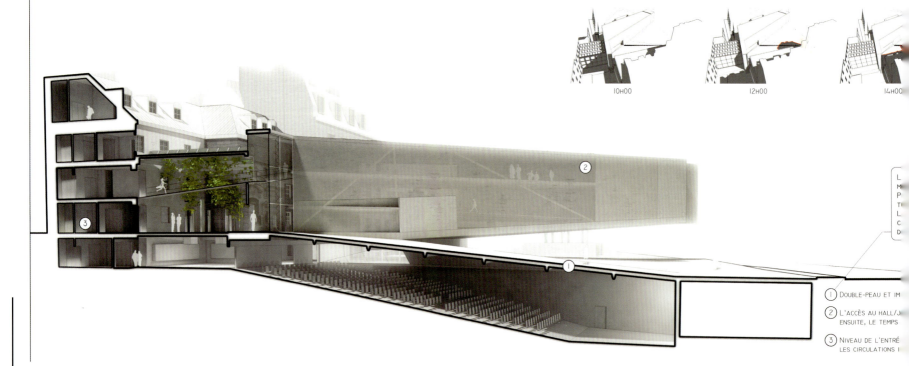

1 DOUBLE-PEAU ET IM

2 L'ACCÈS AU HALL/J
ENSUITE, LE TEMPS

3 NIVEAU DE L'ENTRÉ
LES CIRCULATIONS I

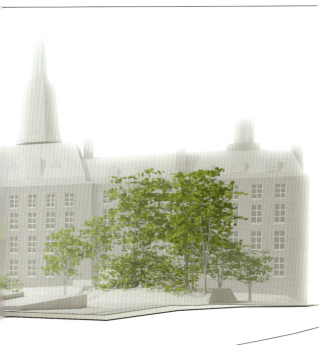

嘉奖项目

多米尼克·彭色列（加拿大，蒙特利尔）

设计师选择运用多层结构，项目的分叉结构与雕塑家理查德·塞拉的作品有异曲同工之妙。该项目在建筑上的大胆表达注定会引起争议。——编者

本页和对页
竞赛展示板图片

Citation

Dominique Poncelet
Montréal, canada

The author chose a multi-layered solution to accommodate the program, whereby the pronged nature of the structure is very reminiscent of the works of the sculptor, Richard Serra. This is a strong statement in architectural expression that would most certainly raise controversy. -Ed.

THIS PAGE AND OPPOSITE
Images from competition boards

RAFA JENN - MONA LISA (2009)

▶ IMPLANTATION

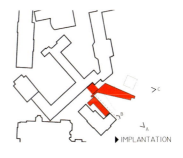

16H00

▶ FAÇADE DOUBLE-PEAU ET PARVIS :
   POTENTIEL SOLAIRE AUX ÉQUINOXES

A DOUBLE-PEAU PRÉSENTE UNE IMPRESSION SUR VERRE QUI REPREND LA TEXTURE BLANCHE ET
E ET DU SÉMINAIRE.
CETTE PAROI DEMEURE LÉGÈREMENT TRANSLUCIDE DEPUIS L'EXTÉRIEUR MAIS CONSERVE SA
'INTÉRIEUR.
E C'EST L'ÉCLAIRAGE INTÉRIEUR QUI DEVIENT LE PLUS PUISSANT, LE RÉSULTAT EST INVERSÉ ET
TÉRIEUR QUE LA PAROI EST TRANSPARENTE, LAISSANT APPARAÎTRE L'ORGANISATION SPATIALE

") LONGE DEUX ORGANES PUBLICS (COMME POUR L'ENTRÉE ACTUELLE). LE PARCOURS PRÉSENTE
IR HISTORIQUE DE L'ENTRÉE "CÔTE DE LA FABRIQUE" (VOIR PLAN PERSPECTIVE DU NIVEAU 2).

E" : CETTE MISE AU MÊME NIVEAU DES DEUX ENTRÉES SIMPLIFIE

AA060

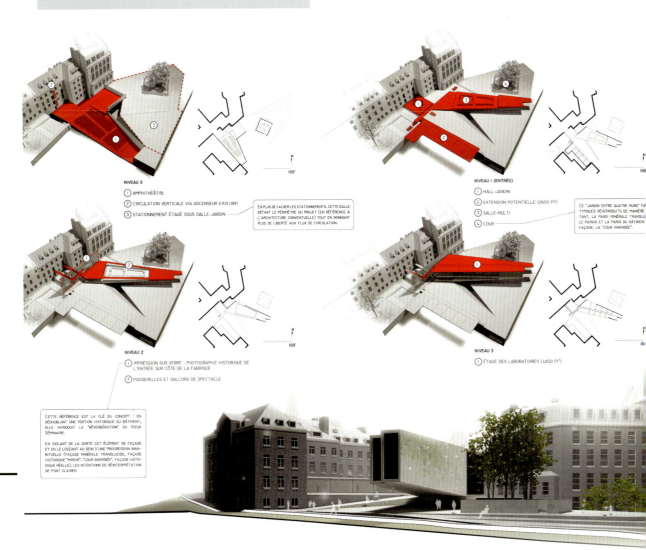

NIVEAU 0
① AMPHITHÉÂTRE
② CIRCULATION VERTICALE VIA ASCENSEUR EXISTANT
③ STATIONNEMENT ÉTAGÉ SOUS DALLE-JARDIN

EN PLUS DE CACHER LES STATIONNEMENTS, CETTE DALLE DÉFINIT LE PÉRIMÈTRE DU PROJET (EN RÉFÉRENCE À L'ARCHITECTURE CONVENTUELLE) TOUT EN DONNANT PLUS DE LIBERTÉ AUX FLUX DE CIRCULATION.

NIVEAU 1 (ENTRÉE)
① HALL-JARDIN
② EXTENSION POTENTIELLE (2650 PI²)
③ SALLE-MULTI
④ COUR

CE "JARDIN ENTRE QUATRE MURS" F...
TYPIQUES RÉINTRODUITS DE MANIÈRE
TANT, LA PAROI MINÉRALE TRANSLU...
LE PARVIS ET LA PAROI DU BÂTIMENT
FAÇADE, LA "COUR INVERSÉE"

NIVEAU 2
① IMPRESSION SUR VERRE : PHOTOGRAPHIE HISTORIQUE DE L'ENTRÉE SUR CÔTE DE LA FABRIQUE
② PASSERELLES ET BALCONS DE SPECTACLE

NIVEAU 3
① ÉTAGE DES LABORATOIRES (4650 PI²)

CETTE RÉFÉRENCE EST LA CLÉ DU CONCEPT : EN DÉDOUBLANT UNE PORTION HISTORIQUE DU BÂTIMENT, ELLE INTRODUIT LA "RÉVERBÉRATION" DU VIEUX SÉMINAIRE.

EN ISOLANT DE LA SORTE CET ÉLÉMENT DE FAÇADE ET EN LE LOGEANT AU SEIN D'UNE PROGRESSION INHA-BITUELLE (FAÇADE MINÉRALE TRANSLUCIDE, FAÇADE HISTORIQUE "MIROIR", "COUR INVERSÉE", FAÇADE HISTO-RIQUE RÉELLE), LES INTENTIONS DE RÉINTERPRÉTATION SE FONT CLAIRES.

A Community Icon Revisited

复兴的社区地标

印第安纳波利斯纪念广场竞赛

The Indianapolis Monument Circle Competition

The Arc de Triomphe, Brandenburg Gate, Red Square, and even the Washington Monument all share a common theme: they represent the spirit of their cities in the most visual urbanistic and symbolic sense. Always located in the central core, they may fulfill different functions—a traffic mode, gateway to the old city, or just central gathering place—but without such symbols, those cities would lose more than part of a historic past. Many small towns in the U.S. have their own courthouse squares; but few can rival Indianapolis' Monument Circle. By virtue of its central downtown location, high visibility, and historic landmark status, it occupies a special chapter in the urban history of our nation.

Monument Circle as we know it today was the subject of an 1887-88 international design competition intended to commemorate veterans of the Civil War. Attracting 70 entries, including five from foreign countries, it was won by the Prussian, Bruno Schmitz. According to the jury, Schmitz's entry, title "Symbol of Indianapolis," was "on the whole the most striking and brilliant of all those presented." Rather than sitting isolated within Circle Park, Schmitz's design engulfed the entire site, cascading down to the sidewalk line in layers of stone terraces, staircases, fountains and groups of sculpture. The design was praised for its holistic treatment of the site and monument as one unified composition. In an inversion of the memorial hall scheme, Circle Park was treated as an outdoor memorial space, a sort of inside-out panorama of memorial sculpture (For more about that competition, go to: www.ratioblog.com/?p=3016).

After the Memorial was completed in 1903, it became a focal point of city activity. Although completed in an era of the horse-drawn carriage, the appearance of the automobile on the circle has done little to diminish its stature; it is still drawing large crowds, both during working hours and on special occasions. Because of its landmark status, and continuing popularity, some thought it time to rethink the site and its function. At a time when other landmark memorial sites, i.e., the Jefferson National Expansion Memorial (St. Louis Arch), are also undergoing transformative studies, this would also seem to be a logical undertaking, especially with the advent of the automobile. During a conference, CEOs for Cities Livability Challenge in 2010, visiting national experts echoed what so many local leaders had expressed: Monument Circle is extraordinary and merits maximum attention, creativity and vision. Thus the idea for a Monument Circle Idea Competition was born.

凯旋门、布兰登堡门、红场乃至华盛顿纪念碑都享有一个共同的主题：它们通过视觉符号和象征意义展现了城市的精神。它们总是坐落在市中心，具有不同的功能——交通枢纽、古城城门、中心集会场所，但是没有这些象征物，那些城市将失去至少一半的历史底蕴。美国许多小城都有自己的市政广场；但是它们都无法比拟印第安纳波利斯纪念广场。广场的中心位置、良好的能见度和历史底蕴让它在美国城市历史拥有独特的地位。

现有的纪念广场是1887-1888年一次国际设计竞赛的产物，用于纪念美国南北战争。那次竞赛吸引了70支参赛队伍，最终由普鲁士人布鲁诺·史密斯获胜。评审称：史密斯的作品"印第安纳波利斯的象征"是"最具吸引力而闪耀的设计"。史密斯的设计没有与纪念公园分离，而是占据了整个场地，通过一系列的石平台、楼梯、喷泉和雕塑，直达人行道。设计的全盘规划深受赞赏。

纪念碑于1903年建成，成为了城市活动的焦点。尽管广场建于马车时代，汽车的出现丝毫没有减弱它的威风。无论是平时还是节假日，广场每天都吸引大量的人群。出于它的地标地位以及受欢迎程度，人们认为是时候对它进行重建了。在2010年的一次会议中，城市居住挑战协会的执行官们与客座专家们形成了共识：纪念广场是一个非凡的场所，值得我们以最大的创造力和注意力来设计。由此，诞生了纪念广场概念竞赛。

FUTURE LAND USE MAP - DOWNTOWN AREA

This map illustrates proposed future land uses of the Mile Square by 2020, as determined during the Regional Center Plan 2020 planning process. This proposes an increase in high-density residential projects in several areas within a half-mile walk of the Circle.

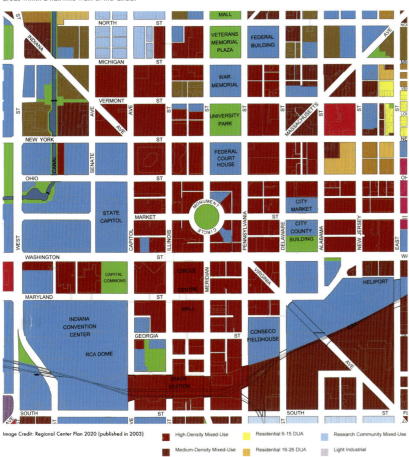

Image Credit: Regional Center Plan 2020 (published in 2003)

■ High-Density Mixed-Use	□ Residential 6-15 DUA	■ Research Community Mixed-Use
■ Medium-Density Mixed-Use	□ Residential 16-26 DUA	□ Light Industrial
■ Non-Core Commercial	□ Residential 27-49 DUA	■ Heavy Industrial
□ Non-Core Office	■ Residential 50+ DUA	■ Parks and Open Space
■ Core Support	□ Research and Technology	■ Public and Semi-Public

The jury chose first, second and third place winners. A "people's choice" winner was chosen by virtue of a ballot vote by local citizens. The premiated designs in order of their ranking were:

First Place ($5,000)
From Inertia to Inner-circle
Jean-Baptiste Cueille / François David
Paris, France

Second Prize ($2,000)
Centering Indianapolis
Tom Gallagher, Ben Ross and Brian Staresnick, Ratio Architects
Indianapolis

Third Prize ($1,000)
Nexus: Indianapolis
Studio Three Architects - Brian Hollars, Lohren Deeg and Kerry LaPrees
Muncie, Indiana

People's Choice ($1,000)
Greg Meckstroth
Philadelphia, PA

评审们选出了一、二、三等奖获得者。经过当地市民的投票，产生了"最受欢迎奖"。获奖设计分别为：
一等奖（5,000美元）
从惯性到内环
让－巴蒂斯特·库埃勒/弗朗西斯·大卫（法国，巴黎）
二等奖（2,000美元）
印第安纳波利斯中心化
汤姆·加拉格尔；本·罗斯和布里安·斯泰尔斯尼克；比例建筑事务所（印第安纳波利斯）
三等奖（1,000美元）
连结：印第安纳波利斯
3号工作室建筑事务所——布里安·霍拉尔斯；劳伦·迪格和拉里·拉普利斯（印第安纳州，曼西）
最受欢迎奖（1,000美元）
格雷格·麦克斯特罗斯（宾夕法尼亚州，费城）

moveable pavilions

Threshold - clay pavement

Terraces - red brick

red brick (existent)

On-street parking & plants

Lawn

Green pavement

Water miror & water games

Fountains (existent)

First Place

Jean-Baptiste Cueille
Francois David
Paris, France

The first-place team presented a scheme, which at least psychologically reduces the distance from the buildings at the outer edge to the inner Circle. Their idea is to transform the circulation pattern so that the inner part of the Circle becomes more activity friendly. They looked at the grid system and proposed a visual cross with the monument at the center to visually facilitate this idea. Additional water and vegetation were also suggested as a way to provide better areas for small gatherings. In the words of the designers, "Contained inside a ring, the original form creates a underutilized space. We propose to liberate the monument from its circular shape to inscribe it in a 'Cross.' This new shape clearly inserts the column in a downtown scale, alongside Market Street and Meridian Street. The Cross shape matches both the width of the stairs & basins and the surrounding streets, while still respectful of the solemnity of this -important landmark." This proposal would seem to solve some of the problems inherent in locating a landmark inside a traffic circle by suggesting more connectivity from the outer edge to the inner circle. -Ed

一等奖

让－巴蒂斯特·库埃勒；弗朗西斯·大卫（法国，巴黎）

一等奖团队的设计在心理上最大限度地缩减了建筑外围和内部转盘之间的距离。他们的理念改造了路径结构，让转盘变得更加活跃。他们着眼于坐标系统，利用中央纪念碑打造了一个视觉交叉点。附加的水景设施和绿化植物为小型集会提供了更好的场所。根据设计师所说，"转盘中的原始造型没有被充分利用。我们计划将纪念广场于转盘中解放，将它改造成一个'交叉点'。"——编者

Built Context: We suggest emphasizing the center by creating guidelines that would extend the limits of each building, increasing a *local scale*. These will generate stronger tangent lines to the interior circle.

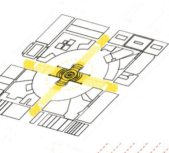

Monument: We propose to liberate the monument from its circular shape to inscribe it in a "Cross". This new shape clearly inserts the column in a *downtown scale*, alongside Market Street and Meridian Street.

Grid: the Circle has a critical geographical meaning. As the city and state's center, we propose to introduce a reduced representation of the city-grid inside the circle. This third layer of guidelines brings back a representation of the *City scale* to its place of origin.

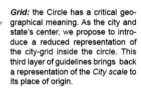

These *3 layers* are the *inspiration for a new design for the Circle*, evocating 3 scales from the smallest to the largest.

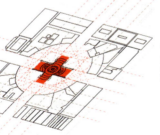

New spaces appear and, as they *overlap, interactive relationships* are created.

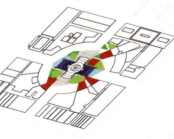

A range of *moveable pavilions* offer venues for a variety of activities (food, shopping, information, events). This active elements move according to *special events* and *sunlighting*, generate *potential spaces*.

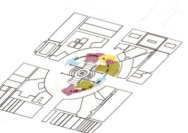

THIS PAGE AND OPPOSITE
Elements from competition boards
BELOW
Section
OPPOSITE, ABOVE
Pedestrian view to monument

本页和对页
竞赛展示板元素
下图
剖面图
对页，上图
从行人视角看纪念碑

SIDEWALK

ROAD
slow conduction zone

TERRACE GARDEN
for public & customers

THRESHOLD
Huge open space for tourists
& events

SOLDIERS' AND SAILORS' MONUMENT

PODIUM
restaurant, outdoor bar &
place for special events

PAVILLION
tourist information, gift shop
hot-dog & drinks

SIDEWALK
pedestrian circulation and privative
slow-conduction for the Hilton Hotel

INERTIA / Monument Circle is characterized by its structure. Activities settle on the edges, while center is only used in touristic/contemplative ways. The concentric system, from building facades to the Monument, generates inertia.

INNER CIRCLE/
We want to let people –use and circulate around the circle more freely– By generating new links between center and edge, we introduce a dynamic system creating new venues and spaces.

IDEA No. 211
PROVIDE PLACES TO LIVE
FOR MIXED INCOME LEVELS
BY RENOVATING UNDER UTILIZED
OFFICE SPACE AND NEW CONSTRUCTION

IDEA No. 57
MAXIMIZE STREETFRONT RETAIL SPACES
TO INCREASE FOOT TRAFFIC THAT CREATES
A BUSTLING STREETSCAPE THROUGHOUT
THE ENTIRE DAY

IDEA No. 979
DESIGN FOR FUTURE GENERATIONS
USE SUSTAINABLE MATERIALS AND
UTILIZE GREEN TECHNOLOGY

IDEA No. 39
EAT LOCAL
PROVIDE A GROCERY STORE FILLED
WITH LOCALLY GROWN PRODUCE

IDEA No. 1880
STEAL A LANE
OF TRAFFIC TO CREATE SPACES
FOR OUTDOOR CAFES, AND VENDORS

IDEA No. 201
GIVE PEOPLE PLACES TO SIT
THAT VARY FROM MOVEABLE CHAIRS,
STEPS, BENCHES, WALLS, LAWNS.

IDEA No. 77
SHARE THE STREET
WITH CARS, DELIVERY TRUCKS,
CYCLISTS, AND PEDESTRIANS

IDEA No. 689
PROVIDE A PARK PAVILLION
WITH A PLACE TO EAT, DRINK,
SIT, AND RELAX

IDEA No. 12
SPLASH IN THE FOUNTAIN
ALLOW PEOPLE TO COOL OFF
AND PLAY IN THE WATER

RECOMMENDATION No. 1XX

CREATE SCENARIOS FOR FLEXIBLE SPACES THAT
ACCOMODATE A VARIETY OF PROGRAMMED EVENTS
THROUGHOUT THE YEAR

二等奖

汤姆·加拉格尔；本·罗斯和布里安·斯泰尔斯尼克
比例建筑事务所（印第安纳波利斯）

该设计团队提出了若干提升步行区的方案，最激进的是从建筑旁边的转盘中"窃取一
条行车道"供行人使用。这包含了拓展零售空间至广场区域。此外，他们还提议封闭
转盘上的部分交通，用于举办农家市场和当地庆典等活动。——编者

Second place

Tom Gallagher, Ben Ross and Brian Staresnick

Ratio Architects, Indianapolis

This team proposed a number of interventions to increase pedestrian traffic, the most radical being "stealing a traffic lane" on the building side of the circle to promote a people place. This would include extending retail farther into the circle area. Additionally, they proposed closing off some of the traffic on the circle to accommodate events such as a farmer's market and local celebrations. In Europe, such closings-off of traffic and turning streets into pedestrian malls is possible. In this country, even partially rerouting traffic away from retail areas would no doubt run up against some formidable opposition—especially since our experience with pedestrian malls in this country has had very mixed results. -Ed

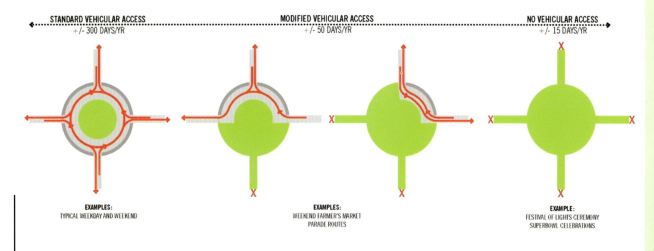

STANDARD VEHICULAR ACCESS
+/- 300 DAYS/YR

MODIFIED VEHICULAR ACCESS
+/- 50 DAYS/YR

NO VEHICULAR ACCESS
+/- 15 DAYS/YR

EXAMPLES:
TYPICAL WEEKDAY AND WEEKEND

EXAMPLES:
WEEKEND FARMER'S MARKET
PARADE ROUTES

EXAMPLE:
FESTIVAL OF LIGHTS CEREMONY
SUPERBOWL CELEBRATIONS

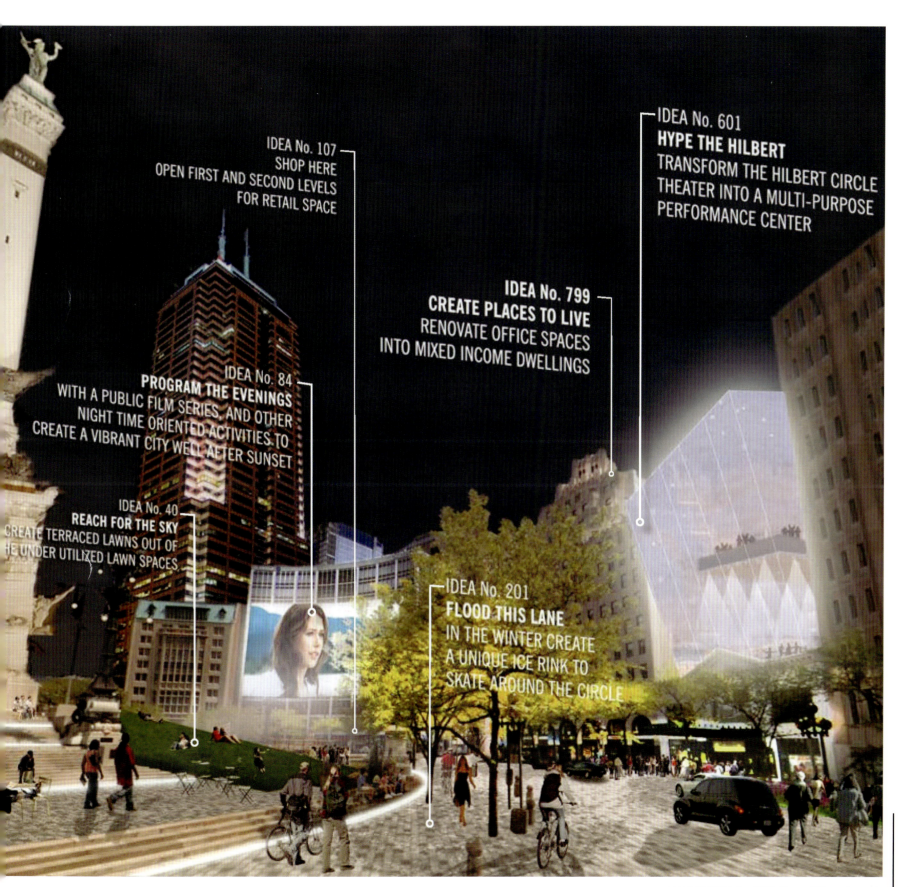

IDEA No. 601
HYPE THE HILBERT
TRANSFORM THE HILBERT CIRCLE THEATER INTO A MULTI-PURPOSE PERFORMANCE CENTER

IDEA No. 107
SHOP HERE
OPEN FIRST AND SECOND LEVELS FOR RETAIL SPACE

IDEA No. 799
CREATE PLACES TO LIVE
RENOVATE OFFICE SPACES INTO MIXED INCOME DWELLINGS

IDEA No. 84
PROGRAM THE EVENINGS
WITH A PUBLIC FILM SERIES, AND OTHER NIGHT TIME ORIENTED ACTIVITIES TO CREATE A VIBRANT CITY WELL AFTER SUNSET

IDEA No. 40
REACH FOR THE SKY
CREATE TERRACED LAWNS OUT OF THE UNDER UTILIZED LAWN SPACES

IDEA No. 201
FLOOD THIS LANE
IN THE WINTER CREATE A UNIQUE ICE RINK TO SKATE AROUND THE CIRCLE

NEXUS : INDIANAPOLIS | **05.** / ENCLOSURE

the trees' broadleaf enclosure frames the Monument to give it a clear visual backdrop

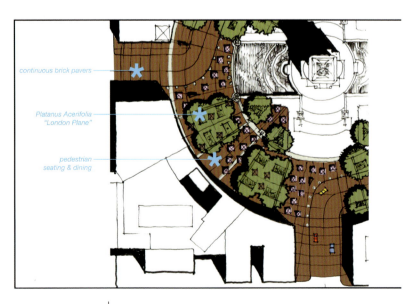

continuous brick pavers

Platanus Acerifolia "London Plane"

pedestrian seating & dining

NEXUS : INDIANAPOLIS | **04.** / SOUTHWEST QUADRANT PLAN - PHASE 1

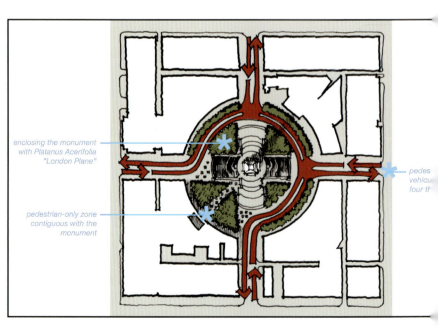

enclosing the monument with Platanus Acerifolia "London Plane"

pedestrian-only zone contiguous with the monument

pedes vehicu four th

NEXUS : INDIANAPOLIS | **02.** / SITE PLAN

Third Place

Studio Three
Team
Lohren Deeg
Kerry Laprees
Brian Hollars
Ball State University
Muncie, Indiana

The third-place team, Studio Three Architects, also had to be well acquainted with the site, as Muncie's Ball State University maintains a local presence in downtown Indianapolis. They proposed a rather radical, but doable plan to isolate the southwest quadrant of the site—closing off that section to vehicles—to turn it into a pedestrian-only zone. To enhance the view of the monument, they suggested planting a tree species, Platanus Acerifolia, also known as the 'London Plane,' which can be found in Bryant Park and elsewhere.

Justifying a new traffic pattern, the team stated, "This design proposes that the SOUTHWEST QUADRANT of the circle be defined as a contiguous pedestrian zone with that of the monument, and retaining full vehicular circulation on all four "spokes." The circulation associated with the Arts Garden, Circle Centre Mall, Convention Center, Stadium, and Illinois Street all occurs to the southeast of the circle. IMAGINE THE ABILITY OF THE VISITOR TO FIND TRUE DESTINATION VALUE in the form of pedestrian oriented businesses, outdoor beverage and dining, and the joy of human contact. This portion of the circle also benefits from the most SHADE during warm summer days, adding vitality to the circle in the evening hours in a defined urban space." -Ed

east / west

三等奖

3位建筑事务所
团队
布里安·霍拉尔斯；劳伦·迪格和拉里·拉普利斯（鲍尔州立大学；印第安纳州，曼西）

获得三等奖的设计团队充分熟悉了场地。他们提出了一个相当激进而可行的方案——通过封闭机动车道，将广场的西南角隔开，使其变成步行区。为了提升纪念碑的景色，他们提议种植法国梧桐。团队打造了全新的交通模式，"这一设计将西南角打造成了一个连续的步行区域，并且保留了四面的车辆交通。"——编者

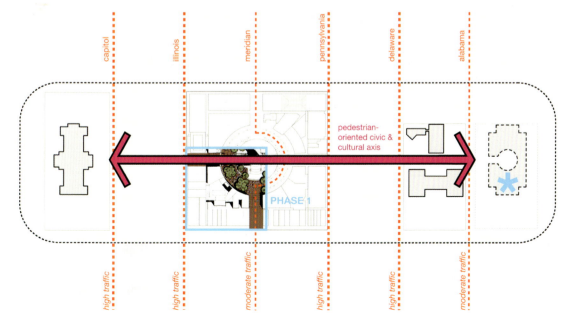

capitol | illinois | meridian | pennsylvania | delaware | alabama

pedestrian-oriented civic & cultural axis

PHASE 1

high traffic | high traffic | moderate traffic | high traffic | high traffic | moderate traffic

circle representative section

new south meridian st. pedestrian plaza
looking north

MoCo Coffee

TAVOLO

circle implementation phasing

phase 1

ACTIVATE SPACE

initiate Monument Circle management group to spearhead space programming and bring more events to Circle.

activate ground floor retail spaces and encourage outdoor dining to bring added life to the street.

encourage unique signage and awnings.

encourage significant ground floor fenestration to connect outside space with indoor activity.

phase 2

HARDSCAPE RESTORATION

improve alleyways with sustainable infrastructure, increased lighting, and connections where possible.

improve existing red brick throughout Governor's Square and create tasteful hardscape pattern to exemplify uniqueness.

phase 3

CREATE CONNECTIONS

north. green connection to University Park via landscaped medians.

south. pedestrian connection to S. Meridian St. and Circle Centre Mall via plazas and outdoor seating.

east. visual connection to City Market via newly installed Market Street arches.

west. visual connection to State Capitol via newly installed Market Street arches.

phase 4

SOFTSCAPE ENHANCEMENT

improve existing green space on south side of Circle by eliminating chained fence barriers.

extend green space into existing auto travel lanes, creating pedestrian only space on south side of Circle.

improve green space on north side of Circle with increased plantings, floral displays, and art.

circle design concept

CLARIFY the EDGE

green connection

visual connection

visual connection

pedestrian connection

MEANINGFUL CONNECTIONS

GREEN RELIEF

PEDESTRIAN REVIVAL

circle perspective

before

after northwest quadrant

circle perspective

after looking southeast

before

**People's Choice
Greg Meckstroth**
Philadelphia, PA

If we didn't know any different-ly, we might assume that this was a real estate agent bent on selling some property. The 'before' and 'after' presentation is a strategy that certainly would appeal to the layperson. This proposal would seem to be primarily a facelift, with lots of street furniture pro-viding the site with some needed ambience.

Apparently, this strategy is one targeted for a summer audience — in other words, limited to about six months of the year before inclement weather sets in.

Although this entry may not exhibit some of the more innova-tive features of those preferred by the jury, it is perfectly clear why the public leaned toward this scheme as a favorite. In cases like this where the public is asked to vote their preferences, it can often give us a clue about a city's cul-tural background. -Ed

最受欢迎奖
格雷格·麦克斯特罗斯（宾夕法尼亚州，费城）

如果我们不明真相，可能会以为该项目是地产商在售楼。"前后对比"的策略对外行特别具有吸引力。该项目就像一次整容手术，众多的街道设施为场地营造了所需的氛围。显然，这个设计的受众是夏天的游客，也就是严冬之前的六个月。尽管设计没有呈现出吸引评审的创意特征，它完美地吸引了公众。在这种最受欢迎的投票中，我们总是一瞥城市文化背景。

Turning a Wasteland into a Community

将废墟变为社区

The Gowanus Lowline Ideas Competition

by Dan Madryga

The landscape of waste: it is a common feature in any big city. Left in the wake of decentralized cities and waning industry, the neglected postindustrial terrain is an unavoidable blemish on the built environment. The desolate, ugly, contaminated vestiges of abandoned factories, overstuffed trash dumps and discontinued mills were pushed out of site and out of mind for decades as Americans sought refuge in suburbia. Yet as urban centers are gradually redeveloped and society expresses increased concern about environmental crises, these harmful, marginalized sites are becoming more difficult to ignore. On Brooklyn's doorstep lies one such wastescape: the dormant and noxious Gowanus Canal. With help from the recent Gowanus Lowline ideas competition, locals are beginning to seriously contemplate a restorative future for this type of ailing urban environment.

郭瓦纳斯运河下游概念竞赛

丹·马德里加

废墟：这是大城市中常有的景象。这些在城市分散化和工业衰退的背景下后工业化区域是建成环境中的污点。当美国人在郊区寻找新天地的时候，荒凉、丑陋的废弃工厂，杂乱的垃圾堆和废弃的制造厂已经被遗忘多年。但是，在城市中心正在经历重新开放和社会对环境问题日益关注的背景下，这些废墟变得让人难以忽视。布鲁克林的边缘就有这样一片废墟：隐蔽而肮脏的郭瓦纳斯运河。郭瓦纳斯运河下游概念竞赛让当地居民开始严肃考虑此类城市环境问题的未来解决方案。

It is not surprising that the Gowanus Canal and its environs constitute one of the few underdeveloped areas of increasingly vibrant Brooklyn, New York. Constructed in the mid 19th century as a key transportation and commercial link between Brooklyn and New York City, the banks of the canal quickly became host to a toxic assortment of gas plants, paint factories, mills, and coal yards. The ensuing decades took their environmental toll and left the waterway brimming with poisonous discharges of industry. The result is a murky, malodorous, literally lifeless waterway that continues to pose serious environmental and health threats long after the factories and mills were boarded up. The EPA has deemed the canal "one of the nation's most extensively contaminated water bodies," and in March 2010 added the site to the Superfund National Priorities List to facilitate extensive investigation into its problems.

While the federal government invests its own interest in the troubled site, a steady increase in local efforts has paved the way for the Gowanus Lowline Competition. This open, international ideas competition was organized by Gowanus by Design, a non-profit organization founded by local architects David Briggs and Anthony Deen and committed to seeking creative and restorative design solutions that point to a brighter future for the canal and its neighboring community. The first of a planned series of canal-centric design exercises, Gowanus by Design's inaugural competition, subtitled "Connections," revolves around a single question: "How do you define connection(s) relative to the Gowanus Canal and secondly, how is this understanding realized?"

Although this simple premise may seem a bit general, the physical, demographical, and ecological circumstances of the canal site essentially speak for themselves in establishing a design challenge: balancing environmental remediation with the creation of a livable community connected to its urban environs is a top priority. And by keeping the program open-ended, Gowanus by Design attracted a diverse a range of ideas that will hopefully spark an engaging dialogue concerning the future possibilities of this post-industrial landscape.

Gowanus by Design attracted a diverse a range of ideas that will hopefully spark an engaging dialogue concerning the future possibilities of this post-industrial landscape.

郭瓦纳斯设计组织吸引了各种各样的设计概念，为这个后工业区域带来了充满希望的前景。

不难理解，郭瓦纳斯运河和它的周边地区是生机勃勃的布鲁克林地区仅有的不发达区域之一。运河建于19世纪中期。作为连接布鲁克林和纽约城区的主要交通商业连接线，运河两岸迅速发展成为天然气厂、油漆工厂、制造厂和煤场的混合区域。数十年的工业发展对环境造成了极大的破坏，污染了水道。在工厂和制造厂倒闭多年之后，其所遗留的环境问题仍然困扰着水道。美国环境保护局认为该运河是"美国污染最严重的水体之一"，并于2010年3月将该地区列入"超级基金国家优先处理区域名单"，以便于对其问题进行调查研究。

在联邦政府投入资金的同时，当地人也努力为竞赛铺路。这次开放式国际概念竞赛由郭瓦纳斯设计组织（由当地建筑师大卫·布里格斯和安东尼·迪恩所创建的非盈利组织，旨在寻求运河及其周边地区未来发展的创意和修复设计方案）举办。郭瓦纳斯设计组织的第一次竞赛副标题为"连接"，围绕着一个问题展开："如何定义郭瓦纳斯运河的连接价值并实现它？"

尽管这个简单的前提看似宽泛，运河区域的地理、人口以及生态环境还是对设计提出了挑战：平衡环境修复与活跃社区之间的关系是重中之重。开放式项目设计让郭瓦纳斯设计组织吸引了各种各样的设计概念，为这个后工业区域带来了充满希望的前景。

LEFT
Gowanus Canal, present condition

左图
郭瓦纳斯运河现状

THIS PAGE
Gowanus site diagrams by First Place team

本页
郭瓦纳斯场地图解，由一等奖团队提供

ESTUARY
BOUNDARY

PARK

HARDSCAPE

WETLANDS

PARKING

VACANT

POLLUTED

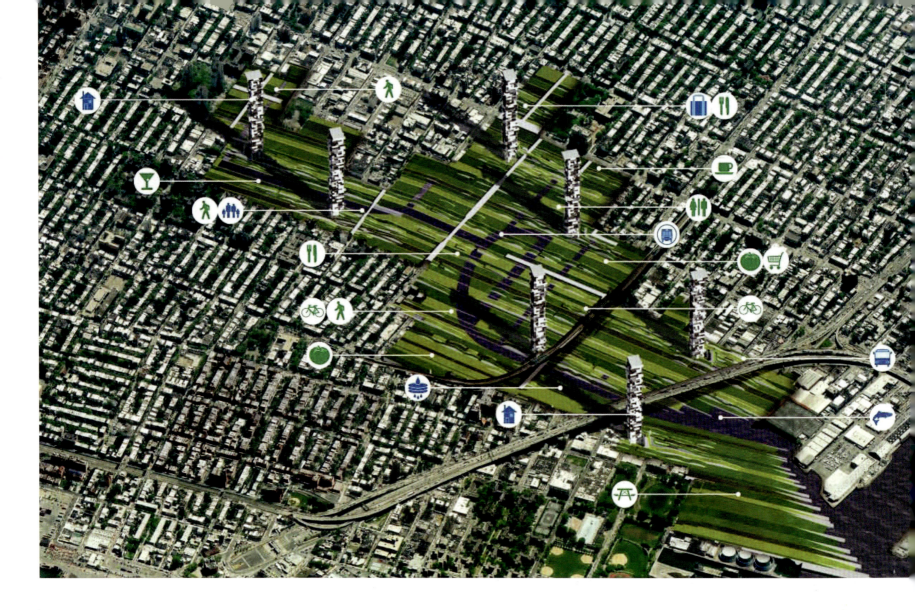

First Place

"Gowanus Flowlands"
Tyler Caine
Luke Carnahan
Ryan Doyle
Brandon Specketer
New York, NY

THIS PAGE AND OPPOSITE
Competition board illustrations and diagrams

一等奖

"郭瓦纳斯流畅区域"
泰勒·凯因；卢克·卡尔纳罕；赖安·多利；
布兰登·斯贝克特尔（纽约）

本页和对页
竞赛展示板示意图表

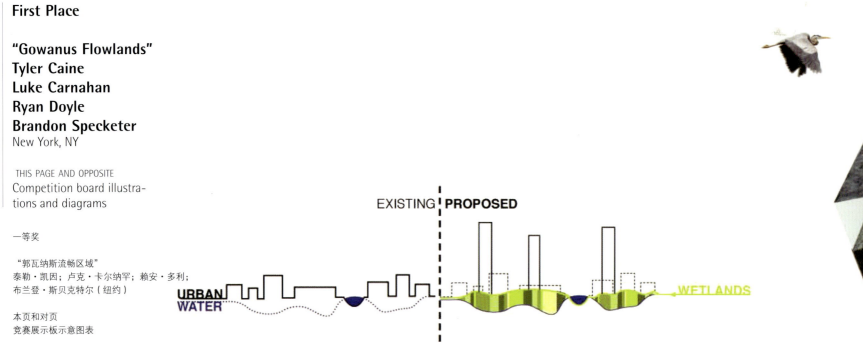

EXISTING | PROPOSED

URBAN WATER

WETLANDS

First Place

The winning team, hailing from New York, operated on the premise that a successful intervention must address dual problems of the Gowanus Canal: environmental contamination and urban depopulation. "A truly sustainable urban ecosystem depends on creating a vibrant and walkable community that is both ecologically and economically self-supportive. The success of the neighborhood relies on this pair of systems being spatially and functionally complementary to one another instead of traditionally being considered in opposition."

The team has essentially created an urban wetland – a complex, well-detailed filtration ecosystem seamlessly interwoven with a network of pedestrian circulation paths, recreational spaces, and retail and commercial buildings tucked underneath vegetated roofs. The jury was pleased with this unique integration, noting that it "presents a more compelling urban condition and suggests a new type of urban landscape that suggests living with remediation."

一等奖

优胜团队来自纽约，认为成功的规划必须着重郭瓦纳斯运河两个方面的问题：环境污染和城市人口下降。"一个真正的可持续城市生态系统依赖于打造一个活跃而适于步行的社区，这个社区必须兼具生态和经济自主性。社区的成功需要这两个系统相互依存，而不是相互独立。"

该团队在根本上打造了一个城市湿地———一个复杂而精密的过滤生态系统与步行流通区域、休闲空间以及带有绿色屋顶的零售和商业建筑紧密结合。评审盛赞这个独特的整合规划，认为它"呈现了更具吸引力的城市环境并且引入了一个全新类型的城市景观"。

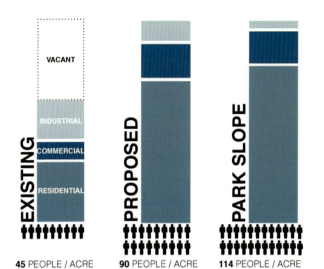

VACANT

INDUSTRIAL

COMMERCIAL

RESIDENTIAL

EXISTING — 45 PEOPLE / ACRE

PROPOSED — 90 PEOPLE / ACRE

PARK SLOPE — 114 PEOPLE / ACRE

FINAL TREATMENT LEVEL OF SHALLOW OYSTER BEDS OVER SAND AGGREGATE IN GRASS PATCHES FOR WATERBORNE CHEMICAL REMOVAL.

SECONDARY SURFACE FLOW FREE WATER BASIN. AQUATIC REED PLANTS IN VEGETATED SUBMERGED BED FOR DISTILLED PURIFICATION. COMPLEX UNDERWATER ENVIRONMENT SUPPORTING VARIED BIRD SPECIES.

PRELIMINARY OPEN WATER, CATCHMENT, AND RETENTION BASIN. SUBMERGED AQUATIC PLANTS OVER A LAYER OF ORGANIC SAND COMPOSITE. FOSTERS SHALLOW POND HABITAT.

LOW-HEIGHT PLANT GROUPS IN ARID SOIL FOR SUB-SURFACE STORMWATER FILTRATION. WHEAT AND BUFFALO GRASSES FOR HYDROCARBON REMOVAL.

GRAVEL-BASED DRY SOIL PRE-TREATMENT WITH POPLAR TREES FOR EMBEDDED HEAVY METAL REMOVAL FROM SOIL.

Second Place

"[f]lowline"
Aptum/Landscape Intelligence

Gail Fulton
Roger Hubeli
Julie Larsen
University of Illinois, Urbana

RIGHT
Competition board

The second place team wisely posits a phased design framework that would gradually develop over time as the canal is slowly remediated. Barges would transport waste material generated from canal dredging to create "Pooling Parks," a system of bioswales creating artificial ecologies long the banks. "Floating Forest Barges" would be stationed along the canal, each holding various social, recreational, and entertainment functions. The jury appreciated the team's "clever strategy" and "sophisticated, realistic approach to phasing. [It] adapts and responds to changing conditions and offers a vision of the future."

二等奖

"流线"
阿普托姆/景观设计事务所
盖尔·富尔顿；罗杰·胡贝利；朱莉·拉尔森（乌尔班纳，伊利诺伊大学）

右图
竞赛展示板

二等奖团队巧妙地设想了一个阶段性设计框架，将随着运河的缓慢修复逐步完成。驳船将清除运河中的废料，打造"水池公园"——一系列生态沼泽将沿着河岸打造人造生态环境。"漂浮的森林驳船"将停驻在运河上，各具社交、娱乐和休闲功能。评审欣赏他们的"奇思妙想"和"纯熟、现实的阶段性规划。它适应并反映了不断变化的环境并且打造了良好的前景。"

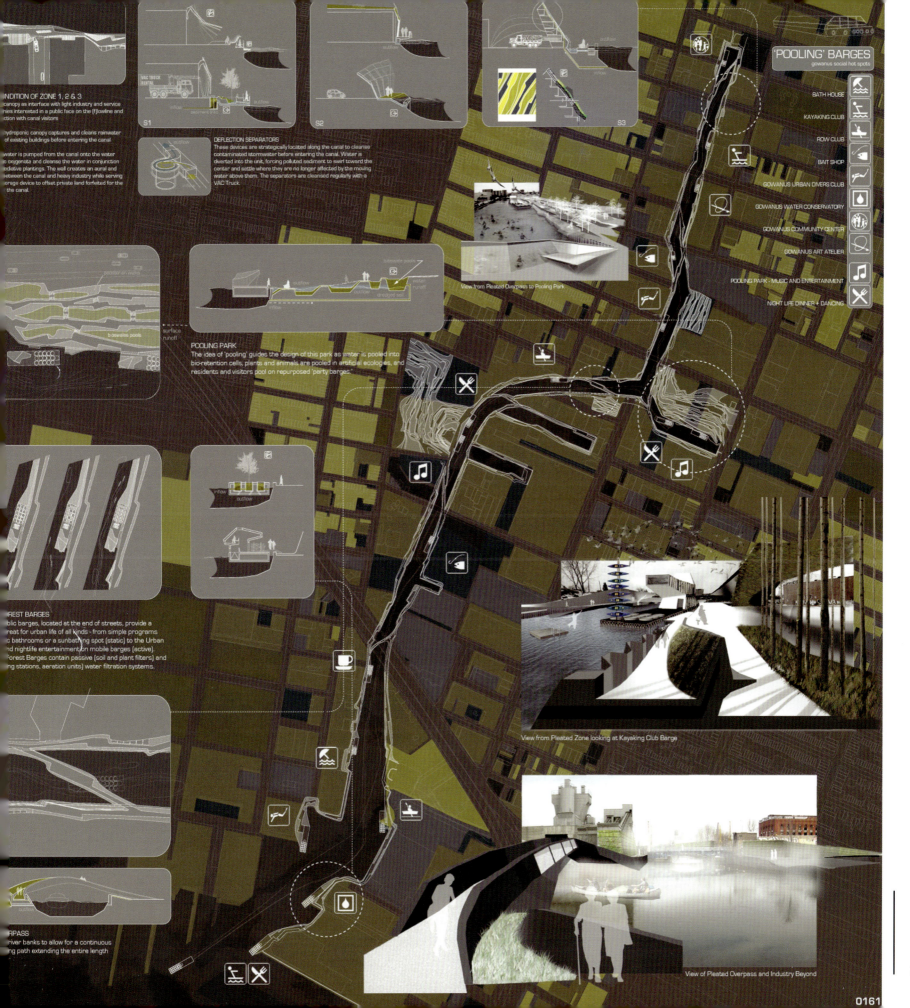

CONDITION OF ZONE 1, 2 & 3
canopy as interface with light industry and service
nies interested in a public face on the [f]lowline and
ction with canal visitors

Hydroponic canopy captures and cleans rainwater
of existing buildings before entering the canal.

water is pumped from the canal onto the water
n oxygenate and cleanse the water in conjunction
ediative plantings. The wall creates an aural and
between the canal and heavy industry while serving
orage device to offset private land forfeited for
the canal.

DEFLECTION SEPARATORS
These devices are strategically located along the canal to cleanse
contaminated stormwater before entering the canal. Water is
diverted into the unit, forcing polluted sediment to swirl toward the
center and settle where they are no longer affected by the moving
water above them. The separators are cleansed regularly with a
VAC Truck.

S1　　S2　　S3

POOLING PARK
The idea of 'pooling' guides the design of this park as water is pooled into
bio-retention cells, plants and animals are pooled in artificial ecologies, and
residents and visitors pool on repurposed 'party barges.'

'POOLING' BARGES
gowanus social hot spots

BATH HOUSE
KAYAKING CLUB
ROW CLUB
BAIT SHOP
GOWANUS URBAN DIVERS CLUB
GOWANUS WATER CONSERVATORY
GOWANUS COMMUNITY CENTER
GOWANUS ART ATELIER
POOLING PARK - MUSIC AND ENTERTAINMENT
NIGHT LIFE DINNER + DANCING

View from Pleated Overpass to Pooling Park

REST BARGES
blic barges, located at the end of streets, provide a
reat for urban life of all kinds - from simple programs
c bathrooms or a sunbathing spot [static] to the Urban
d nightlife entertainment on mobile barges [active].
orest Barges contain passive [soil and plant filters] and
ng stations, aeration units) water filtration systems.

View from Pleated Zone looking at Kayaking Club Barge

RPASS
river banks to allow for a continuous
ng path extending the entire length

View of Pleated Overpass and Industry Beyond

Flush Basin Curtain

Future canal basin contamination is limited by flushing water through a variety of remediation methods including floating islands where exposed roots uptake nutrients, provide cover and food for fish, and a substrate for beneficial microbe colonization, phytoremediation terraces that cleanse CSO water before it enters the canal, and a curtain wall made from custom fabricated architectural grill by a local industry that filters all site run off through biomatrix and phytoremediation layers.

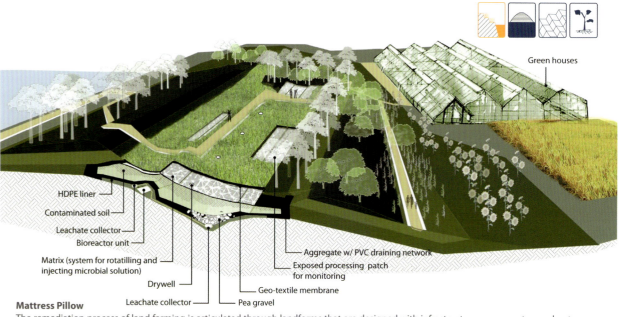

Mattress Pillow

The remediation process of land farming is articulated through landforms that are designed with infrastructure necessary to accelerate the soil cleansing process. Sod attached to a geo-textile membrane provides a movable surface of containment over layers of contaminated soil and layers of bio-augmented material. A network of tubing connected to vertical drywells provides a means for monitoring contamination, maintaining moisture levels, and injecting the biomatrix mixture specific to the soil being treated. This network also affords a means of mechanically mixing the soil until it is clean and ready for removal. This intensive remediation system can eventually be used for more moderate methods of soil health maintenance related to phytoremediation and urban agriculture.

REGISTRATION #: 0061

Honorable Mention

"Domestic Laundry" Agergroup

Jessica Leete
Claire Ji Kim
Shan Shan Lu
Winnie Lai
Albert Chung

Boston, Massachusetts

THIS PAGE AND OPPOSITE
Images from Competition board

In "Domestic Laundry," a three-phase clean-up and development plan utilizes a number of well-researched bioremediation techniques for a realistic yet inspiring solution.

荣誉奖

"内部清洗"
阿格尔集团

杰西卡·利特；克莱尔·吉·金；姗姗·卢；维尼·赖；艾伯特·钟（马萨诸塞州，波士顿）

本页和对页
竞赛展示板图片

在"内部清洗"设计中，一个三期的清理和开发规划利用一系列生物修复技术来实行了实际而鼓舞人心的解决方案。

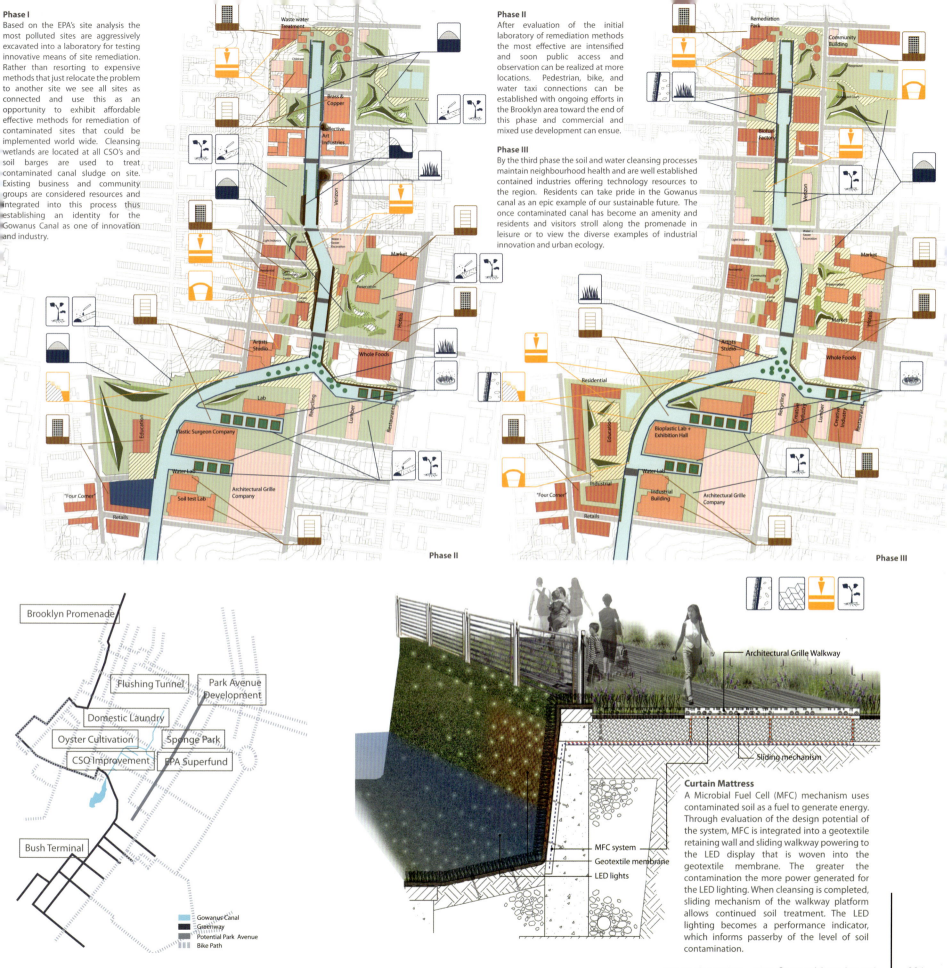

Phase I

Based on the EPA's site analysis the most polluted sites are aggressively excavated into a laboratory for testing innovative means of site remediation. Rather than resorting to expensive methods that just relocate the problem to another site we see all sites as connected and use this as an opportunity to exhibit affordable effective methods for remediation of contaminated sites that could be implemented world wide. Cleansing wetlands are located at all CSO's and soil barges are used to treat contaminated canal sludge on site. Existing business and community groups are considered resources and integrated into this process thus establishing an identity for the Gowanus Canal as one of innovation and industry.

Phase II

After evaluation of the initial laboratory of remediation methods the most effective are intensified and soon public access and observation can be realized at more locations. Pedestrian, bike, and water taxi connections can be established with ongoing efforts in the Brooklyn area toward the end of this phase and commercial and mixed use development can ensue.

Phase III

By the third phase the soil and water cleansing processes maintain neighbourhood health and are well established contained industries offering technology resources to the region. Residents can take pride in the Gowanus canal as an epic example of our sustainable future. The once contaminated canal has become an amenity and residents and visitors stroll along the promenade in leisure or to view the diverse examples of industrial innovation and urban ecology.

Curtain Mattress

A Microbial Fuel Cell (MFC) mechanism uses contaminated soil as a fuel to generate energy. Through evaluation of the design potential of the system, MFC is integrated into a geotextile retaining wall and sliding walkway powering to the LED display that is woven into the geotextile membrane. The greater the contamination the more power generated for the LED lighting. When cleansing is completed, sliding mechanism of the walkway platform allows continued soil treatment. The LED lighting becomes a performance indicator, which informs passerby of the level of soil contamination.

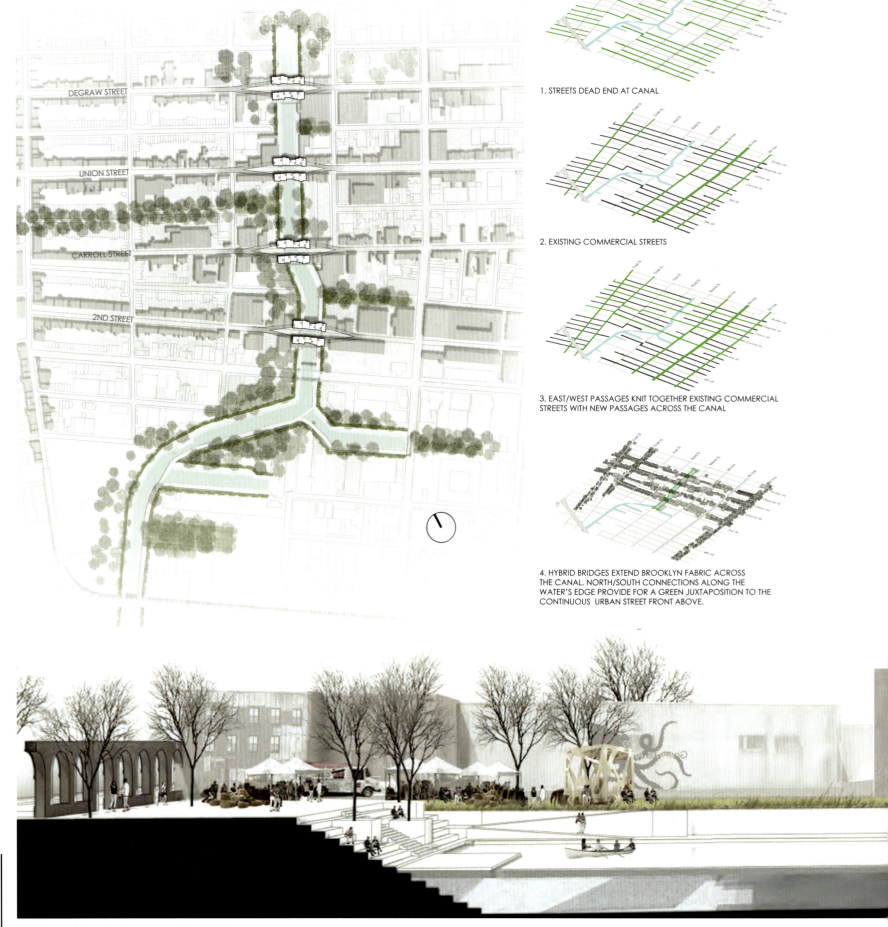

DEGRAW STREET

UNION STREET

CARROLL STREET

2ND STREET

1. STREETS DEAD END AT CANAL

2. EXISTING COMMERCIAL STREETS

3. EAST/WEST PASSAGES KNIT TOGETHER EXISTING COMMERCIAL STREETS WITH NEW PASSAGES ACROSS THE CANAL

4. HYBRID BRIDGES EXTEND BROOKLYN FABRIC ACROSS THE CANAL. NORTH/SOUTH CONNECTIONS ALONG THE WATER'S EDGE PROVIDE FOR A GREEN JUXTAPOSITION TO THE CONTINUOUS URBAN STREET FRONT ABOVE.

Honorable Mention

"Made in Brooklyn: Bridges for Local Artisans & Industry"
Nathan Rich
Miriam Peterson
Brooklyn, NY

荣誉奖

"布鲁克林制造：当地工匠和工厂的桥梁"
南森·里奇；米里亚姆·彼得森（纽约，布鲁克林）

THIS PAGE
Images from Competition board

本页
竞赛展示板图片

"Made in Brooklyn," the only top choice entry that does not deal explicitly with ecological recovery nevertheless provides an intriguing community development concept: bridges lined with commercial, industrial, and residential functions would help catalyze urban development along the canal and better connect the adjacent neighborhoods.

"布鲁克林制造"是唯一一个没有明确处理生态修复问题的获奖作品。然而，它提供了一个有趣的社区发展概念：连接商业、工业和住宅功能的桥梁将帮助促进运河区域的城市发展，更好地连接周边区域。

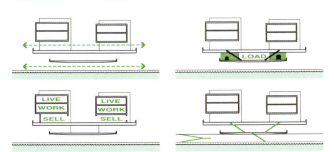

SECTIONAL CONNECTIONS

ID # 0080

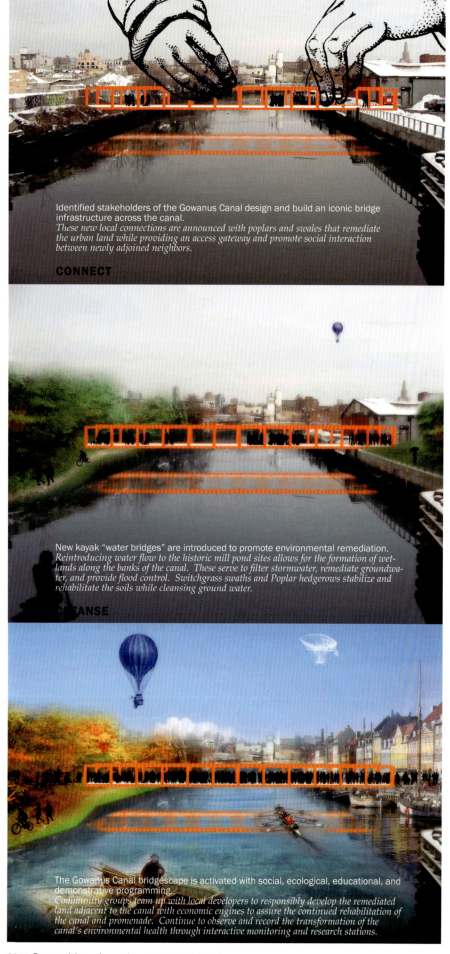

Identified stakeholders of the Gowanus Canal design and build an iconic bridge infrastructure across the canal.
These new local connections are announced with poplars and swales that remediate the urban land while providing an access gateway and promote social interaction between newly adjoined neighbors.

CONNECT

New kayak "water bridges" are introduced to promote environmental remediation.
Reintroducing water flow to the historic mill pond sites allows for the formation of wetlands along the banks of the canal. These serve to filter stormwater, remediate groundwater, and provide flood control. Switchgrass swaths and Poplar hedgerows stabilize and rehabilitate the soils while cleansing ground water.

CLEANSE

The Gowanus Canal bridgescape is activated with social, ecological, educational, and demonstrative programming.
Community groups team up with local developers to responsibly develop the remediated land adjacent to the canal with economic engines to assure the continued rehabilitation of the canal and promenade. Continue to observe and record the transformation of the canal's environmental health through interactive monitoring and research stations.

Gowanus Low Line
B.Y.O.B.
BuildYourOwnBridge

A community design exercise for remediating urban landscapes and promoting local stewardship

Honorable Mention

Austin+Mergold LLC
Jason Austin
Alex Mergold
Jessica Brown
Sally Reynolds
Philadelphia, PA

THIS PAGE AND OPPOSITE
Images from Competition board

本页和对页
竞赛展示板图片

"Build Your Own Bridge" proposes a "bridge-scape" that helps connect two separated Brooklyn neighborhoods and is set within a network of recreation and remediation parks.

荣誉奖

奥斯丁+梅尔格德设计公司
杰森·奥斯丁；亚历克斯·梅尔格德；杰西卡·布朗；莎莉·雷诺兹（宾夕法尼亚州，费城）

"打造你自己的桥梁"提出了一个"桥景"概念，将布鲁克林两个独立的街区连接起来，并于其中嵌入了休闲和修复公园。

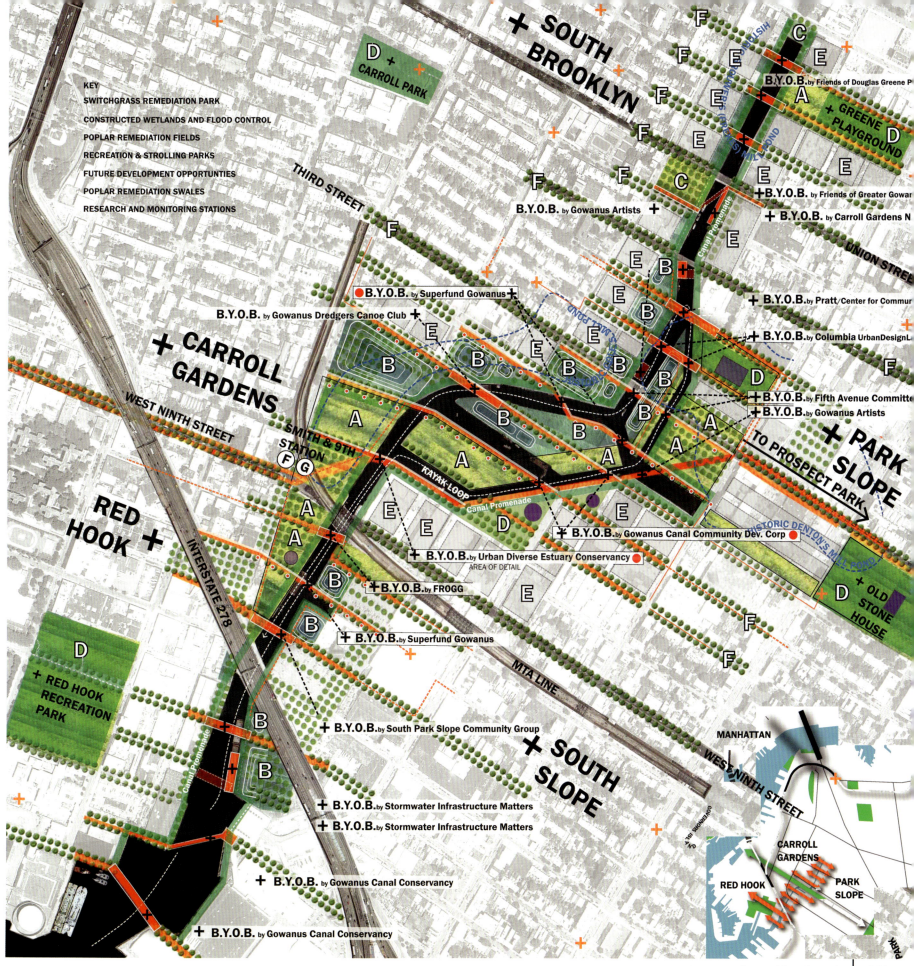

KEY

SWITCHGRASS REMEDIATION PARK

CONSTRUCTED WETLANDS AND FLOOD CONTROL

POPLAR REMEDIATION FIELDS

RECREATION & STROLLING PARKS

FUTURE DEVELOPMENT OPPORTUNITIES

POPLAR REMEDIATION SWALES

RESEARCH AND MONITORING STATIONS

+ SOUTH BROOKLYN

D + CARROLL PARK

B.Y.O.B. by Friends of Douglas Greene P

+ GREENE PLAYGROUND

B.Y.O.B. by Friends of Greater Gowar

B.Y.O.B. by Carroll Gardens N

THIRD STREET

B.Y.O.B. by Gowanus Artists +

B.Y.O.B. by Superfund Gowanus +

B.Y.O.B. by Gowanus Dredgers Canoe Club +

UNION STREET

B.Y.O.B. by Pratt/Center for Commur

B.Y.O.B. by Columbia UrbanDesignL

+ CARROLL GARDENS

WEST NINTH STREET

SMITH & 9TH STATION F G

B.Y.O.B. by Fifth Avenue Committee

B.Y.O.B. by Gowanus Artists

+ PARK SLOPE

TO PROSPECT PARK

RED HOOK +

INTERSTATE 278

KAYAK LOOP

Canal Promenade

HISTORIC DENTON'S MILL POND

B.Y.O.B. by Gowanus Canal Community Dev. Corp

B.Y.O.B. by Urban Diverse Estuary Conservancy

AREA OF DETAIL

D + OLD STONE HOUSE

+ **B.Y.O.B.** by FROGG

+ **B.Y.O.B.** by Superfund Gowanus

MTA LINE

+ SOUTH SLOPE

D RED HOOK RECREATION PARK

B

+ **B.Y.O.B.** by South Park Slope Community Group

MANHATTAN

WEST NINTH STREET

+ **B.Y.O.B.** by Stormwater Infrastructure Matters

+ **B.Y.O.B.** by Stormwater Infrastructure Matters

CARROLL GARDENS

RED HOOK

PARK SLOPE

+ **B.Y.O.B.** by Gowanus Canal Conservancy

+ **B.Y.O.B.** by Gowanus Canal Conservancy

GOVERNORS ISL.

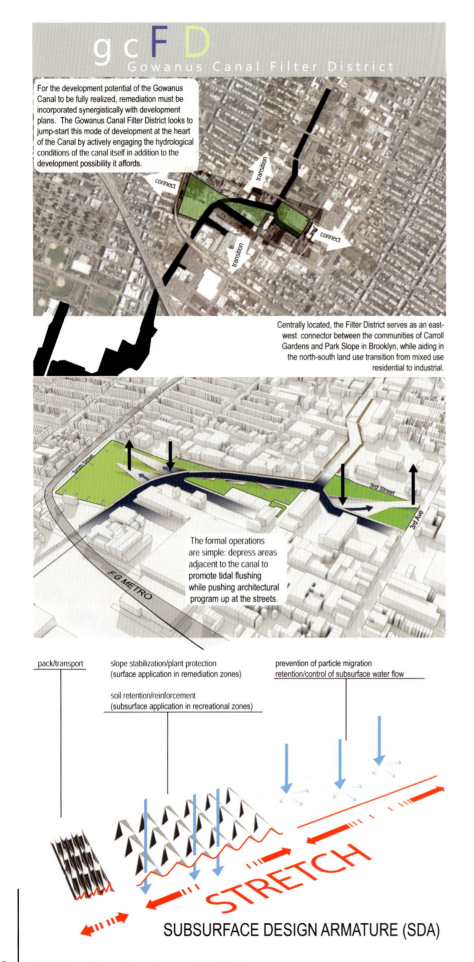

gcFD
Gowanus Canal Filter District

For the development potential of the Gowanus Canal to be fully realized, remediation must be incorporated synergistically with development plans. The Gowanus Canal Filter District looks to jump-start this mode of development at the heart of the Canal by actively engaging the hydrological conditions of the canal itself in addition to the development possibility it affords.

connect

transition

transition

connect

Centrally located, the Filter District serves as an east-west connector between the communities of Carroll Gardens and Park Slope in Brooklyn, while aiding in the north-south land use transition from mixed use residential to industrial.

The formal operations are simple: depress areas adjacent to the canal to promote tidal flushing while pushing architectural program up at the streets.

pack/transport

slope stabilization/plant protection
(surface application in remediation zones)

soil retention/reinforcement
(subsurface application in recreational zones)

prevention of particle migration
retention/control of subsurface water flow

STRETCH

SUBSURFACE DESIGN ARMATURE (SDA)

Honorable Mention

"Gowanus Canal Filter District"
burkholder|salmons
Sean Burkholder
Dylan Salmons
Warren, PA

荣誉奖

"郭瓦纳斯运河过滤区"
布克修德尔I萨尔蒙斯
肖恩・布克修德尔；达兰・萨尔蒙斯（宾夕法尼亚州，沃伦）

本页和对页
竞赛展示板图片

THIS PAGE AND OPPOSITE
Images from Competition board

"郭瓦纳斯运河过滤区"将自然过滤系统与富有魅力的公园融为一体。

"Gowanus Canal Filter District" incorporates natural filtering systems into an attractive park.

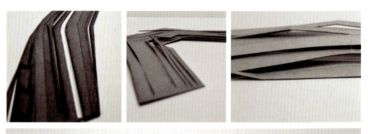

landform/circulation studies

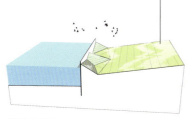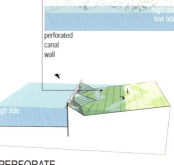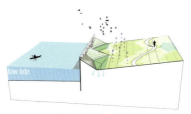

EXISTING CONDITIONS

The existing conditions of many parcels within the site consist of asphalt or concrete pavement, contributing to the excessive stormwater and CSO problem that has plaguing the canal for generations.

EXCAVATE

The initial phase of the site development consists of the removal of soil adjacent to the canal and the installation of the SUBSURFACE DESIGN ARMATURE. This adjustable armature can facilitate soil retention, erosion control and water percolation depending on its installation. The excavated soil will be used on site for the construction of landforms and other development program

PERFORATE

By strategically puncturing the piling along the canal, tidal flushed can extend beyond the wall into the site while still providing protection for barges and ships that continue to use the canal for industrial or recreational purposes.

REMEDIATE/RECREATE

The new protected tidal wetlands will foster remediation of the canal water through both phytoremediative plants and infiltration. These wetlands will additionally provide treatment of stormwater runoff, thus reducing the frequency of combined sewer overflow events.

Miami Seaplane Terminal Competition: A Catalyst for the Watson Island Waterfront

by Eric Goldemberg 迈阿密水上飞机航站楼竞赛：沃森岛滨水区发展的催化剂 埃里克·戈登伯格

Background 背景

DawnTown, an annual architectural ideas competition since 2007, has become a huge success. Attracting over hundred entries from more than 20 countries in 2010, this quest to bring innovative architecture to Downtown Miami has become a staple in the international design community. Its latest installment, the 2010 DawnTown Seaplane Terminal Competition, focused on issues pertaining to waterfront and infrastructural development, and asked the participants to investigate landscape and architectural opportunities that would reenergize the coastal growth of Miami.

Watson Island — the site of this year's competition — holds a dynamic and important position in Biscayne Bay off the coast of Miami. The island was created by land reclamation in 1926 from the dredgings for the shipping channel that provided access to the Port of Miami. It is close to downtown Miami, and can only be accessed by the Macarthur Causeway and Bridge that help comprise highway 395, a major artery for Miami that services Miami Beach. Due to its location, the land has had a history of increasingly more prominent usage. In 1932, Watson Island was considered for the site of Miami's Pan-American Exposition, a World's Fair and "International merchandise mart." Since 1946 it has been home to the Miami Outboard Club and Marina as well as the Miami Yacht Club. The island also housed a seaplane terminal named Chalk's for several decades. In the Summer of 2003, an animal attraction called Jungle Island relocated to the island and two years later, the Miami Children's Museum followed suit. Lately it has been the focus of renewed attention with large investments on infrastructural development as well as a planned luxury hotel complex that includes retail, residences, and a marina for mega yachts. A new project, the Miami Port Tunnel, further anchors the importance of the island and its location by connecting Watson Island to Dodge Island, the home of one of the largest cruise ship and cargo ports in the world. The project will traverse under the shipping channel, and is intended to alleviate the current traffic congestion in downtown.

自2007年以来，黎明镇年度建筑概念竞赛取得了巨大的成就。2010年，竞赛吸引了来自20多个国家的上百件参赛作品，旨在为迈阿密市中心带来创意建筑设计，成为了国际设计界的旗帜。它的最后一部分——2010黎明镇水上飞机航站楼竞赛聚焦于滨水区和基础设施的发展，要求参赛者为迈阿密海岸重塑具有活力的景观和建筑设计。

沃森岛（本年度竞赛的场地）拥有充满活力的重要地理位置，位于迈阿密海岸线上的比斯坎湾。沃森岛形成于1926年的迈阿密码头航道清淤工程。它紧邻迈阿密市区，只能通过麦克阿瑟堤道和大桥进入。得益于自身的地理位置，小岛的地位日益突出。1932年，沃森岛被选为迈阿密泛美展览会——一次国际性商业盛会的场地。自1946年以来，小岛一直是迈阿密小艇俱乐部和迈阿密游艇俱乐部码头的所在地。沃森岛还用一个名为"粉笔"的水上飞机航站楼。2003年夏，丛林岛动物乐园落地沃森岛。两年之后，迈阿密儿童博物馆紧随其后，也来到了沃森岛。最近，它成为了大型基础设施和一家奢华酒店投资的焦点。酒店将包含零售、住宅和巨型游艇码头。新项目迈阿密隧道进一步凸显了沃森岛的重要性，连接了沃森岛和道奇岛。项目将贯穿航道，缓解市中心的交通拥堵。

一等奖

"迈阿密林中通道"
CA景观设计事务所

特雷弗·柯尔提斯+希尔维亚·金（韩国，
首尔）

下图
航站楼鸟瞰图

First Place

"Miami Glades"
CA Landscape
Trevor Curtis + Sylvia Kim
Seoul, South Korea

BELOW
Birdseye view of terminal site

THICKET STRUCTURE

A thicket of knotted mangrove roots and stems grow out into the water to colonize the coastal area. Despite exposure to air, roots are strongly embedded into the thick muddy ground to withstand blistering hurricanes.

In view of these factors, structural columns for the terminal are designed in response to the hurricane-prone climate of Miami and to accommodate sustainable features such as rainwater harvest and solar energy capture. This structure will be further used as an aquarium, greenhouse, and resting area to create an innovative and attractive ambiance.

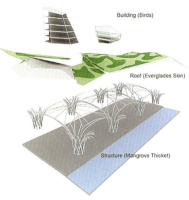

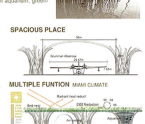

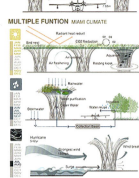

EVERGLADES SKIN

An in-depth analysis of Everglades flora was made to compose the planting plan for the Everglades garden. Curvilinear forms, found in the Everglades, were incorporated to develop natural and orderly pedestrian circulation to match the overall flow of Watson Island. From the seaplane, the pedestrian trail will create a miniature landscape model of the Everglades. The masterplan blends sustainable architecture organically with verdant vegetation, and places the terminal as an ecological focal point and addresses the idea "park in the city". The Everglades which has halted extensive expansion of the city will bring new perspective to visitors and local inhabitants.

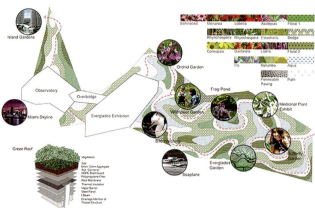

PROGRAM

The design for the DawnTown Seaplane Terminal creates a sustainable icon for Watson Island. Programs have been made accordingly and space was planned to accommodate administrative and services of the terminal. The buildings take form from the steel thicket-structure and reinforced glass.

First and second floors of the terminal building are for both international and domestic terminals. Service circulation is placed nearby to provide convenient access to terminal programs. The control room is placed on the highest elevation to capture the fine views of managing safe take-offs and landings.

Bridged to a separate structure, the restaurant building is planned for flexible accommodation. Since the terminal has limited access after working hours, the restaurant building has its own operational plans. It is designed to hold a wide range of events.

Circulation from the two buildings leads to the hangar roof. The design concept of the Everglades garden is to create a tranquil and contextually rich environment for visitors by reflecting elements characteristic of the Miami's Everglades landscape. Visitors can rest and enjoy this green space by taking a stroll, jogging, socializing, relaxing, and various other activities.

Thicket-inspired structures are designed to withstand the subtropical climate of Miami by integrating cooling system and water recycling system. This will raise ecological awareness as well as visually attract visitors' attention. This masterplan aims to establish a thrilling transit business and offer a thrilling experience to all users.

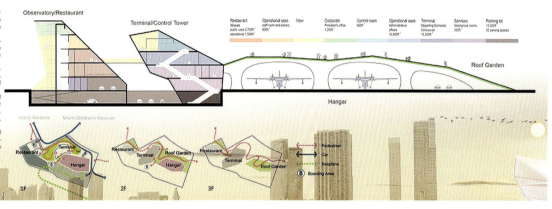

CONCEPT

Birds

Miami is an eco city. The natural environment of the Everglades provides fertile habitat for a variety of species, especially wading birds. Since these birds are ranked higher up in the food chain, it can be inferred that there is an abundance of primary producers (e.g. sawgrass) and secondary producers (e.g. water insects). In other words, the vigorous micro-ecosystem distinguished by sawgrass and the biodiversity within mangrove thickets proves that the Everglades represent a healthy ecosystem. The design concept of the terminal and restaurant structures takes the form of two birds perched on the Everglades. This bird-like structure further reflects the physical and verbal context of Grumman Albatross, representing the aerodynamic aesthetic of a bird-inspired seaplane.

Everglades

At the southern portion of Florida, a subtropical wetland Everglades stretches for an extensive 4000 square miles. Filtering pollutants, it functions as a wildlife haven and the kidney of Planet Earth. Slow in movement, land flows to a stop and keeps connecting and disconnecting to show natural beauty and order. Enveloped by the Everglades patterned vegetation, the ecological roof garden will establish a dramatic view from the sky and accommodate a wide range of events to make the terminal more open and accessible.

Mangroves

Mangroves are trees and shrubs that are found in the tropical and subtropical coastal regions. Many mangrove species have adapted to both saline and freshwater conditions, overcoming the problems of anoxia. In order to intake air, mangroves breathe through root holes that are above water level. The unique root systems of mangroves provide breeding and feeding grounds for a variety of species, particularly marine wildlife. Inspired by the structural stability and biodiversity of mangroves, the design of the terminal's columns highlights these sustainable features.

First Place

"Miami Glades"
CA Landscape
Trevor Curtis + Sylvia Kim
Seoul, South Korea

The project takes the solution of the seaplane terminal to a much larger scale of reference, that of the landscape and the larger community. The building serves its elegant iconic purpose as the culmination of a networked ecology that emphasizes the relationship with nature. As a result this provides a beautiful understanding of the structure by mimicking the unique South Florida Landscape. The roof of the seaplane hangar becomes a natural park offered to the city as a public experience that recreates the wildness of the Everglades, a topography that doubles up as urban event and infrastructural envelope. This large roof is "lifted" from the ground, revealing a structure of bundled tubes that evoke the dense, complex roots of the native vegetation found in the material organization of mangroves and strangler fig trees.
-EG

LEFT, ABOVE
Sections and materials usage
LEFT
Ecology plan
OPPOSITE PAGE
Site comentary

左上
剖面图和材料使用
左图
生态规划
对页
场地解说

一等奖

"迈阿密湿地"
CA景观设计事务所
特雷弗·柯尔提斯+希尔维亚·金（韩国，首尔）

项目将水上飞机航站楼的方案纳入了更大的规划之中，进行了景观和社区规划。建筑以其优雅的造型体现了地标特征，屋顶的网状造型显示了与自然的紧密联系。它的结构完美地诠释了南佛罗里达州景观。水上飞机库上方的屋顶形成了一个自然公园，为城市提供了独特的湿地体验，兼具城市活动和基础外壳功能。这个巨大的屋顶立于地面之上，显示了内部的束管结构，仿佛原生植物复杂的根系一样。——埃里克·戈登伯格

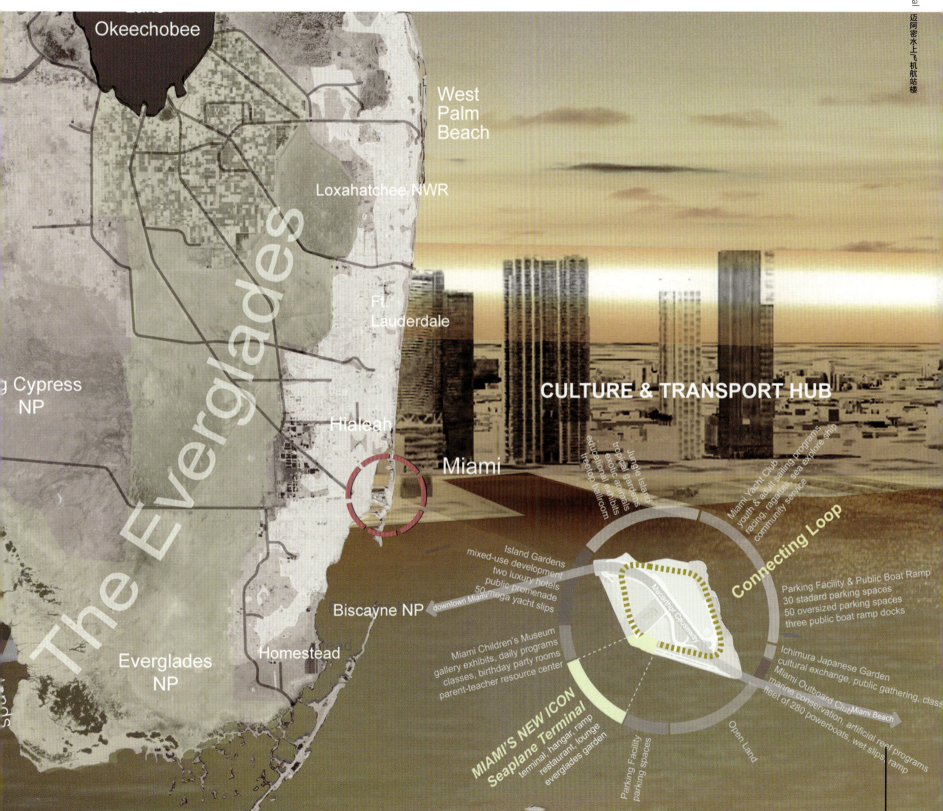

VIEW A: ARRIVAL TO THE AMPHIBIAN SEAPORT

VIEW B: SKY LOUNGE

VIEW C: SITE ORGANIZATION

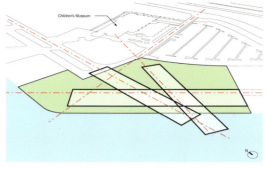

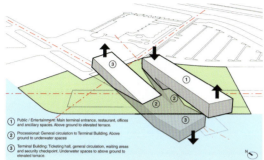

① Public / Entertainment: Main terminal entrance, restaurant, offices and ancillary spaces. Above ground to elevated terrace.

② Processional: General circulation to Terminal Building. Above ground to underwater spaces.

③ Terminal Building: Ticketing hall, general circulation, waiting areas and security checkpoint. Underwater spaces to above ground to elevated terrace.

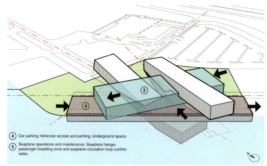

④ Car parking: Vehicular access and parking. Underground space.

⑤ Seaplane operations and maintenance: Seaplane hangar, passenger boarding zone and seaplane circulation loop out/into water.

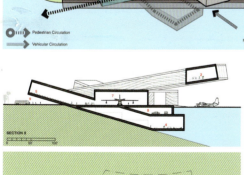

Pedestrian Circulation
Vehicular Circulation

Second Place

"Amphibian"
Stantec
Vicky Chan, Alex Zulas, Liange Otero Colon

New York, NY

二等奖

"水陆两栖"
斯坦泰克
维奇·陈；艾利克斯·朱拉斯；里安格·奥特罗·科隆（纽约）

This entry was chosen for its strong connection to the water. The project takes advantage of it through the plan geometry which juxtaposes four linear components that absorb the predominant direction of privileged visuals and axial alignments of the site. It questions the flatness of the ground level by partially submerging the building along oblique sectional trajectories, creating unprecedented effects for submerged views. It uses the simplicity of its geometry to create a whimsical idea. The project relates to the site very well, and establishes relationships that just seem spectacular, creating fantastic visuals and surprising experiential conditions in terraced interior arrays. -EG

这个设计以其与水的紧密联系而入选。项目在四个并置的线型元素之间充分利用了水元素。他们将建筑沿着一个倾斜的轨道线进行设计，挑战了地平面。它利用简洁的几何造型打造了异想天开的设计。项目与场地十分契合，建立了独一无二的联系，营造了梦幻的视觉效果和奇妙的体验。——埃里克·戈登伯格

VIEW E: CORAL GARDEN

SECTION X

PLAN: UNDER WATER

SECTION Y

PLAN: GROUND LEVEL

SECTION Z

PLAN: ABOVE GROUND

LEGEND

1. GATEWAY (MAIN ENTRY)
2. SKY LOUNGE (RESTAURANT & BAR)
3. CHANNEL (RAMP CIRCULATION)
4. SEA HALL (TICKETING AREA)
5. THE WING (WAITING AREA)
6. HANGAR & SUPPORT SPACES
7. BREEZEWAY DECK (BOARDING AREA)
8. GARAGE
9. DROP OFF AREA
10. CORAL GARDENS

VIEWS (A TO F)
SECTIONS (X TO Z)

43126

DAWNTOWN MIAMI 2010

THE MIAMI LOOP

CONCEPT

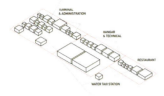

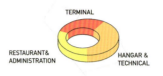

SITE & PROGRAM BREAK DOWN

PROGRAM CONCENTRATED AROUND CENTRAL PLAZA

CLEAR PROGRAM DISTRIBUTION

SHIFT TO ACCOMMODATE TRAFFIC

LIFTING TO MAKE AN ENTRANCE AND VIEWS
THE LOOP!

SITE & CONCEPT

Exposed on Watson island, opposite Miami´s skyline the Miami seaplane terminal
links sky, sea and shore.

This joint function implicates a form without direction, both focused and open. The
MIAMI LOOP is perfectly undirected, and therefore open to arrivals and departures
of all kind.

PROGRAM

By concentrating the program of the terminal around a central plaza a compact
and functional volume is created. By pushing the volume towards the waterfront
all infrastructure can be accomodated.
Lifting the ring upwards creates a gate for planes as well as it allows for views
from the inside of the ring and creates a fantastic panoramic restaurant and view-
ing platform above the waterfront.

TERMINAL & HANGARS

The terminal and hangars are organized on one level around the central plaza,
which results in very functional operational sequences.

RESTAURANT & ADMINISTRATION

The restaurant and the observation platform occupy the most prominent part of
the MIAMI LOOP: the arch cantilevering across the government cut main channel.
The management and admimistraion offices are located in the arch as well. All
functions in the arch are connected by the ` grand stair ´ and elevators

SCALE

Being both attraction and destination, the MIAMI LOOP is a new landmark in central Miami´s skyline.

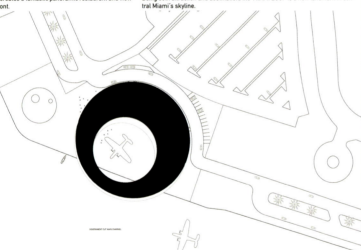

SITEPLAN SC 1/64"=1'-0"

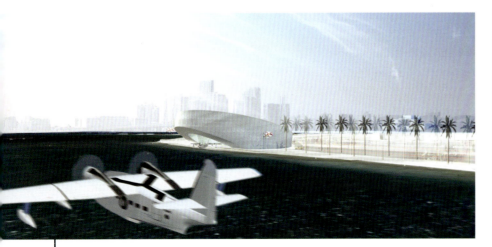

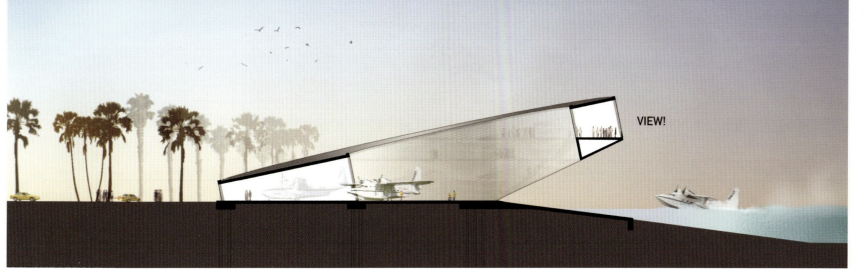

VIEW!

THE MIAMI LOOP

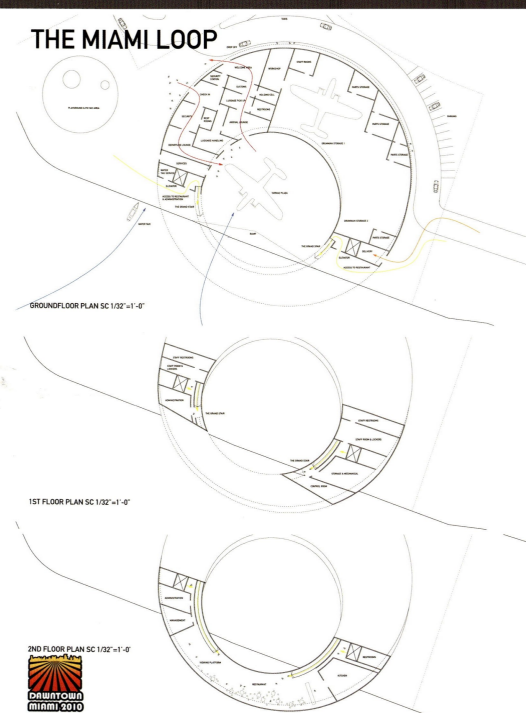

GROUNDFLOOR PLAN SC 1/32"=1'-0"

1ST FLOOR PLAN SC 1/32"=1'-0"

2ND FLOOR PLAN SC 1/32"=1'-0"

Third Place　　　　　　三等奖

"The Miami Loop"　　"迈阿密环"
Gerd Wetzel　　　　吉尔德·维策尔；马丁·普罗
Martin Plock　　　　克（瑞士，巴塞尔）

Basel, Switzerland　　上图，立面图
　　　　　　　　　　　左图，平面图
　　　　　　　　　　　对页，夜景、鸟瞰图和地面视野

The Loop, the third place winner, offers a very strong scheme, a synthetic approach to form that consolidates the diverse programs into a ring-like organization around an open court for boarding. Because of its form, the Loop makes for a great compliment to the American Airlines Arena just across the bay. The building is lifted sectionally in a gesture that facilitates the accessibility of the planes and creates a dramatic relationship of the building in contrast to the surface of the water on one side, and the plateau that it is encrusted into, forming the access plaza. The circular form creates some organizational challenges for the solution of the plans, which are resolved in a tortuous manner and without recognition of the potential advantages of the radial geometry.
-EG

ABOVE
Elevation
LEFT
Floor plans
OPPOSITE PAGE
Night view, perspectives from air and ground

三等奖项目"迈阿密环"提供了一个强有力的规划，将各种不同的功能区结合在一个环形组织结构中。圆环中央的开阔场地用于登机。圆环的造型与海岸对面的美国航空体育馆有异曲同工之妙。建筑提升起来，有利于飞机进出，同时与分列两侧的海水和高原形成了互动。圆环造型在规划上提出了一些挑战，需要进行复杂的解决方案。——埃里克·戈登伯格

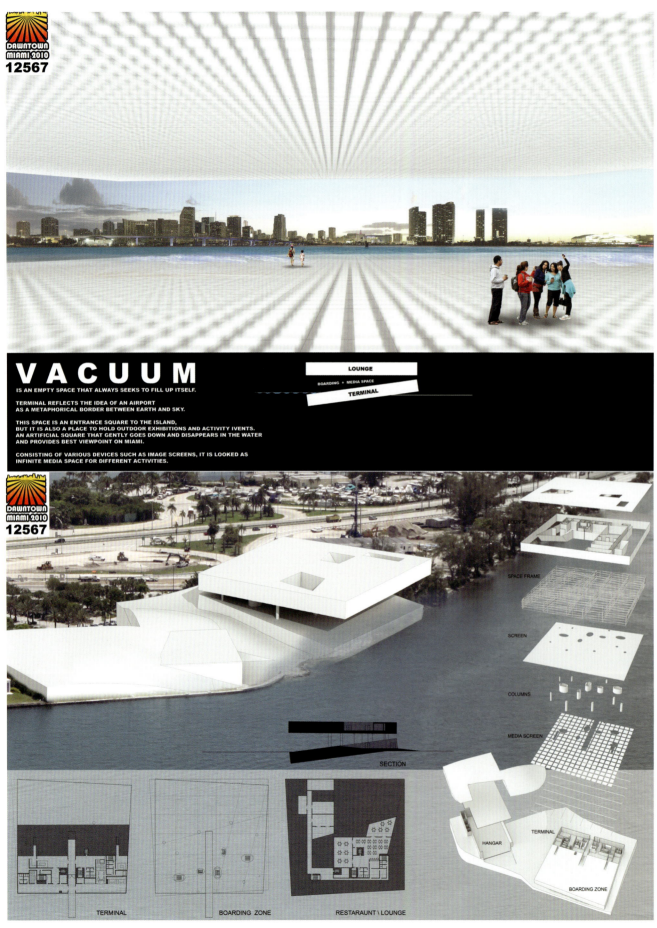

VACUUM

IS AN EMPTY SPACE THAT ALWAYS SEEKS TO FILL UP ITSELF.

TERMINAL REFLECTS THE IDEA OF AN AIRPORT
AS A METAPHORICAL BORDER BETWEEN EARTH AND SKY.

THIS SPACE IS AN ENTRANCE SQUARE TO THE ISLAND,
BUT IT IS ALSO A PLACE TO HOLD OUTDOOR EXHIBITIONS AND ACTIVITY IVENTS.
AN ARTIFICIAL SQUARE THAT GENTLY GOES DOWN AND DISAPPEARS IN THE WATER
AND PROVIDES BEST VIEWPOINT ON MIAMI.

CONSISTING OF VARIOUS DEVICES SUCH AS IMAGE SCREENS, IT IS LOOKED AS
INFINITE MEDIA SPACE FOR DIFFERENT ACTIVITIES.

LOUNGE

BOARDING + MEDIA SPACE

TERMINAL

ROOF

ROOMS

SPACE FRAME

SCREEN

COLUMNS

MEDIA SCREEN

SECTION

HANGAR

TERMINAL

BOARDING ZONE

TERMINAL   BOARDING ZONE   RESTARAUNT \ LOUNGE

1st Honorable Mention

荣誉奖第一名

"Vacuum" "真空"
Nikolay Martynov

Moscow, Russia

尼古拉·马特伊诺夫（俄罗斯，莫斯科）

This entry sets the stage for an ideal appreciation of the Miami skyline underneath a mega-canopy reminiscent of Mies Van der Rohe's "universal space" concept, and a reference to the sublime dimension of the landscape. Most of the program is placed on this floating, thickened canopy which liberates a vast plinth that suffers from certain degree of indeterminacy as well as having a remote connection with the actual movement of the seaplanes. -EG

这件作品完美地呈现了迈阿密的天际线，其巨大的华盖借用了密斯·凡·德罗的"宇宙空间"概念，对景观进行了升华。大多数功能区都设在这个悬浮的增厚华盖之上，从而让巨大的底座缺乏确定感，并且与水上飞机的实际运动也有距离感。——埃里克·戈登伯格

3rd Honorable Mention

荣誉奖第三名

"Large Roof" "大屋顶"
NC-OFFICE NC工作室
Nik Nedev, Peter Nedev, Elizabeth Cardona, Cristina Canton, Jorge San Martin

Miami, Florida

（佛罗里达州，迈阿密）

The authors set up a field condition that articulates a corrugated concrete roof, an echo of other Miami spectacle icons from the 50's and 60's such as the Miami Marine Stadium. It is able to resolve elegantly the necessary large spans of the hangars and it introduces a dark, shadowy atmosphere of brutal use of concrete which gives it a strong expressive character. The interior program argues against the heaviness of the project. However, once one examines the interior and contrasts it to its exterior, the project falls apart.

设计师设计了一个波纹状屋顶，与迈阿密20世纪五六十年代的标志性建筑（例如迈阿密海洋运动场）类似。它巧妙地解决了飞机库的跨度，并且通过混凝土的运用营造出昏暗的氛围，特色鲜明。室内设计则与厚重的外观相反，是项目的败笔。

MIAMI BREATH

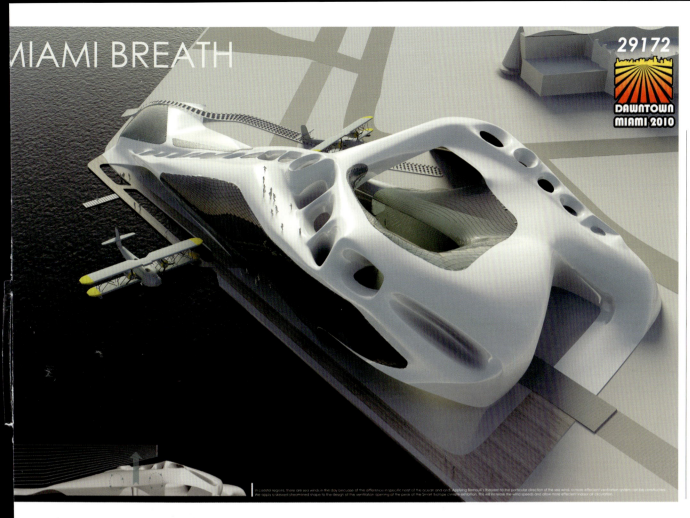

29172
DAWNTOWN MIAMI 2010

荣誉奖第二名

"迈阿密呼吸"
零空间工作室
俊胜·杨；朱赫·金（韩国，首尔）

2nd Honorable Mention

"Miami Breath" Zerovolume
JungSeung Young
Joo Hee Kim

Seoul, South Korea

This proposal creates striking presence on the landscape. The morphology of this terminal evokes the fluidness of the water as well as the erosion and filtering of wind. The porous surfaces indicate very interesting performative possibilities in terms of channeling flows of passengers, although that energy seems misplaced as it only stands as visual allegory to aero-dynamics, a metaphor explored by i conic architectural works such as the TWA Terminal by Eero Saarinen. -EG

设计打造了引人注目的景观。航站楼的形态反映了水的流动性和风的侵蚀过滤作用。多孔的表面在输送客流方面展示了有趣的潜力。项目在形态上比喻了空气动力学，正如埃罗·萨里宁所设计的环球航空公司航站楼一样。

A LARGE STAIR AND RAMP SEQUENCE ESTABLISH AN URBAN EDGE ON THE STREET

AT THE TOP OF THE STAIR/RAMP PEOPLE AWAIT THE ARRIVING PASSENGERS

Notable 2010/2011 Competitions 2010/2011年度知名竞赛

Aberdeen City Garden

3rd Stage Finalists

- Diller Scofidio and Renfro (New York) / Keppie Design (Glasgow) and landscape architect,Olin Studio.
- Foster & Partners (London) / Vladimir Djurovic Landscape Architecture (Beirut) with cost and construction consultant Gardiner & Theobald
- Snøhetta (Oslo) / Gareth Hoskins Architects (Glasgow) working in associatio with engineering and multi-discipline company AECOM
- Gustafson Porter (London) / Niall McLaughlin Architects (London) with urban analyst Space Syntax, engineer Arup and cost adviser Jackson Coles
- Mecanoo Architecten (Delft, Netherlands) / Cooper Cromar (Glasgow) with landscape architect Ian White, engineer Buro Happold
- West 8 urban design & landscape architecture (Rotterdam, Netherlands) / Archial Group (Aberdeen) with engineer Arup

Note: The final decision is between teams 1 and 2. See: http://www.malcolmreading.co.uk/architecturalcompetitions/citygarden/submissions/.

Siemans Headquarters, Munich, Germany

- **Henning Larsen Architects, Copenhagen/Denmark (Winner)**
- Allmann Sattler Wappner Architekten, Munich/Germany
- Auer + Weber + Assoziierte, Munich/Germany
- Baumschlager Eberle, Lochau/Austria
- Hopkins Architects, London/United Kingdom
- Hascher Jehle Architektur, Berlin/Germany
- Ingenhoven architects, Düsseldorf/Germany
- JSWD Architekten, Cologne/Germany
- Léon Wohlhage Wernik Architekten, Berlin/Germany
- Rafael de la Hoz Arquitectos, Madrid/Spain
- Schmidt Hammer Lassen Architects, Aarhus/Denmark
- Schneider Schumacher, Frankfurt am Main/Germany

Victoria and Albert Museum – Exhibition Road

- **Amanda Levete Architects (Winner)**
  London, UK
- Heneghan.peng.architects
  Dublin, Ireland
- Jamie Fobert Architects
  London, UK
- Jun Aoki & Associates
  Osaka, Japan
- Michael Maltzan Architecture (MMA)
  Los Angeles, USA
- Snøhetta & Hoskins
  Oslo/London
- Tony Fretton Architects
  London, UK

Website::www.malcolmreading.co.uk/vanda/shortlist/amanda_levete_architects/

Moving landscapes – Autoroute 20 gateway corridor linking the Montreal-Trudeau Airport to downtown Montreal

- **Brown and Storey Architects, Toronto, Canada**
- **dlandstudio, New York, USA**
- **Gilles Hanicot, Montreal, Canada**

(All three were awarded C$33,000 as first place winners)

Bogata Convention Center, Bogata, Columbia

- **Daniel Bermudez and Juan Herreros (Colombia/Spain) Winner**
- * Snohetta + RIR of New York, USA and Bogotá, Colombia
- * Diller Scofidio + Renfro + UdeB Architecture of New York, USA, and Bogotá, Colombia
- * Saucier + Perrotte Architectes of Montreal, Canada.
- * Zaha Hadid Architects + JMPF of London, England and Medellín, Colombia

Minneapolis Riverfront Design Competition

- **TLS/KVA (Tom Leader/Kennedy Violich), Berkeley/Boston** (Winner)
- Ken Smith Workshop, New York, NY
- Stoss Landscape Urbanism, Boston, Massachusetts
- Turenscape, Beijing, China

Web: http://minneapolisriverfrontdesigncompetition.com/

New Orleans Sustainable Housing Design Competition

- **Judith Kinnard with Tiffany Lin, New Orleans, LA (Winner)**
- Ammar Eloueini Digit-All Studio, New Orleans, LA
- Billes Architecture, New Orleans
- Bild Design, New Orleans
- Mathes Brierre Architects, New Orleans
- Metro Studio, New Orleans
- Trapolin Architects, New Orleans
- Wiznia Architecture + Development, New Orleans

Pylon Design Competition, UK

- **Bystrup Arkitekter, Copenhagen, Denmark (Winner)**
- Ian Ritchie Architects, Jane Wernick Associates and Ann Christopher, Sculptor, London, UK **(Purchase)**
- New Town Studio Engineer: Structure Workshop, Harlow, Essex, UK **(Purchase)**
- Gustafson Porter with Atelier One and Pfisterer, London, UK
- AL_A & Arup, UK
- Knight Architects / Roughan & O'Donovan / ESB International in association with MEGA, UK

Website: http://www.ribapylondesign.com/shortlist

Atlanta History Center Competition

- **Pfeiffer Partners Architects of New York City (Winner)**
- patterhndesign, St. Louis
- MSTSD Inc. / Kallman McKinnell & Wood Architects, Atlanta/Boston
- Stanley Beaman & Sears, Atlanta
- Victor Vines Architecture and Kenneth Hobgood Architects, Raleigh, NC

Changing the Face – Moscow 2011

- **Juan Andres Diaz Parra, Colombia (First Prize)**
- Adrian Reinbroth, Franziska Boettcher and Jenny Grossman, Germany (2nd)
- Joseph Sung, South Korea (3rd Prize)

Museum of the Second World War in Gdansk, Poland

- **Studio Architektoniczne "Kwadrat", Gdynia; Poland Winner**
- Piotr Plaskowicki & partnerzy Architekci, Warszawa; Poland (2nd Prize)
- BETAPLAN S.A., Ateny; Greece (3rd Prize)

Taiwan Tower International Competition

- **First Prize: Sou Fujimoto Architects / Sou Fujimoto, Japan**
  with Fei & Cheng Associates / Philip T.C. Fei, Taiwan
  - Second Prize: soma ZT GmbH / Martin Oberascher, Austria
  with Ricky Liu & Associates Architects+Planners / Ricky Liu, Taiwan
  - Third Prize: Dorin Stefan. Romania
  with DS Birou de Arhitectura, Romania and Lai & Associates, Engineers
  Architects / Chao Chiun Lai, Taiwan
  - Honorable Mention-1: Cook Robotham Arch. Bureau Ltd / Peter Cook, UK
  with Tai Architect & Associates / Tai, Yu-Tse, Taiwan and Buro Happold Ltd., UK
  - Honorable Mention-2: HMC Group / Raymond Pan, USA
  with HOY Architects & Associates / Charles Hsueh, Taiwan

Paris PARC – Paris Sorbonne Université Pierre et Marie Curie Research Center

- **Brjarke Ingels Group I BIG & OFF Architecture (Winner)**
  Copenhagen/Paris with engineers Buro Happold
- MVRDV, Rotterdam, The Netherlands
- Lipsky-Rollet Architectes, Paris
- Mario Cucinella Architects, Bologna, Italy
- Peripherique Architectes, Paris

Taichung Gateway Park International Competiition

- **Catherine Mosbach, France/Philippe Rahm Architectes, France (Winner)**
  with Ricky Liu & Associates, Taiwan
- Stoss, Inc, Boston, USA with Shu Chang Associates Architects, Taiwan (2nd)
- West 8 Urban Design & Landscape / Adriaan Geuze (Netherlands) (3rd)
  with Old Farmer Landscape Architecture Co., (Taiwan)
- Pier Vittorio Aureli, Andrea Branzi Architetto, Favero & Milan Engineering (HM)
  with Tai Architects & Associates, Taiwan

Presidents Park South, Washington, DC

- **Rogers Marvel Architects, New York (Winner)**
- Hood Design Studio, Oakland, California
- Michael Van Valkenburgh Associates (MVVA), Brooklyn, New York
- Reed Hilderbrand Associates, Watertown, Massachusetts
- Sasaki Associates, Watertown, Massachusetts
Website: http://www.ncpc.gov/ppdc/

John Cranko Ballet School, Stuttgart

- *1st Prize* – **Burger · Rudacs Architekten, Munich**
- *2nd Prize* - GMP Generalplanungsgesellschaft mbH, Hamburg
- *3rd Prize* - Nieto Sobejano Arquitectos S.L., Berlin
- *Fourth Prize* - Karl + Probst, Munich
- *Fifth Prize* - Lederer · Ragnarsdóttir · Oei, Stuttgart
- *Citation* - Snøhetta, Oslo
- *Citation* - Delugan-Meissl ZT GmbH, Vienna
- *Citation* - e2a Eckert · Eckert Architekten, Zürich

Airport Lands Design Competition, Edmonton, Alberta

- **Perkins and Will, Vancouver (Winner)**
- Foster and Partners, London, U.K. (Final round)
- KCAP Architects, Rotterdam, The Netherlands (Final round)
- Sweco International AB, Stockholm, Sweden
- BNIM, Kansas City, US

One Prize 2011: Water as the Sixth Borough

- **Ali Fard and Ghazal Jafari, Toronto, Canada (Winner)**
- RUX Design, Long Island City NY (Honorable Mention)
  Russell Greenberg, Christopher Beardsley, and Joseph Corsi
- Cooper Union Institute for Sustainable Design, New York (Honorable Mention)
  Arnold Wu, Kevin Bone, Paul Deppe, Joe Levine, Sunnie Joh, Raye Levine, '
  Al Appleton, and Zulaikha Ayub
- JDKP, USA (Honorable Mention)
  Jeffrey Troutman, Dustin Buck, Kendall Goodman, and Paul McBride

Beton Hala Waterfront Center, Belgrade

- **Team "ARCVS", Belgrade (1st Prize - shared)**
- **Sou Fujimoto Architects with Ove Arup, Tolyo, Japan (1st - shared)**
- Dejan Miljkovi, Jovan Mitrovi, Aleksandar Rodi, Belgrade (3rd Prize - shared)
- MX_SI architectural studio", Belgrade (3rd Prize - shared)

ACSA's 10th Annual Steel Design Student Competition

First Category
- **Will Allport, Nick Barrett, and Jason Butz, Clemson Univ., (1st Prize)**
- Stephen Bonamy and Michael Fontana, Lawrence Technological Univ. (2nd)
Second Category
- Daniel Cesarz, University of Wisconsin-Milwaukee (1st Prize)
- Cody James Glen, Salvador Cabezon, and Maria A. Sanchez Casella (2nd Prize)
  Woodbury University and the University of Buenos Aires
- Dion Dekker, California Polytechnic State University (3rd Prize)
Article by Lee Aviv, graduate student at DAAP, The University of Cincinnati can be
viewed at:: www.competitions.org

London Olympic Park Legacy Design Competition

South Park Competition
- Agence Ter (Paris)
- Gustafson Porter (London)
- James Corner Field Operations
- Ken Smith Landscape Architect
- West 8 (Netherlands)

North Park Competition
- Cottrell &Vermeulen Architecture Ltd (London)
- David Kohn Architects (London)
- erect architecture (London)
- The Landscape Partnership (London)
- Ushida Findlay architects (London)

Photo Credits 图片版权